TRIGONOMETRY

This book gives a simple explanation of the fundamental principles of Trigonometry, its rules and formulae. Only a knowledge of Arithmetic and elementary
Algebra is assumed, Chapter I giving a short resumé of
Geometry and Chapter II a fairly full treatment of logarithms for the reader who has no previous acquaintance
with these subjects. The exercises are carefully graded
and of a practical nature, and metric (S.I.) units are used
throughout.

D1290144

TEACH YOURSELF BOOKS

The book is simply written, bears evidence of long teaching experience, and indeed, is a straightforward account of the foundations of the subject. All important formulae are derived leading up to the solution of triangles, and a useful selection of practical problems follows. Worked examples are given throughout the book. Anyone with some knowledge of arithmetic and elementary algebra and geometry should be able to read the book with profit.

Polytechnic Magazine

TRIGONOMETRY

P. Abbott

TEACH YOURSELF BOOKS
Hodder and Stoughton

First printed 1940
Metricated edition 1970
Twelfth impression 1984

ISBN 0 340 26392 X

Printed in Great Britain for
Hodder and Stoughton Educational,
a division of Hodder and Stoughton Ltd,
Mill Road, Dunton Green, Sevenoaks, Kent,
by Richard Clay (The Chaucer Press) Ltd,
Bungay, Suffolk

INTRODUCTION

Two major difficulties present themselves when a book of this kind is planned.

In the first place those who use it may desire to apply it in a variety of ways and will be concerned with widely different problems to which Trigonometry supplies the solution.

In the second instance the previous mathematical training of its readers will vary considerably.

To the first of these difficulties there can be but one solution. The book can do no more than include those parts which are fundamental and common to the needs of all who require Trigonometry to solve their problems. To attempt to deal with the technical applications of the subject in so many different directions would be impossible within the limits of a small volume. Moreover, students of all kinds would find the book overloaded by the inclusion of matter which, while useful to some, would be unwanted by others.

Where it has been possible and desirable, the bearing of certain sections of the subject upon technical problems has been indicated, but, in general, the book aims at equipping the student so that he will be in a position to apply to his own special problems the principles, rules and formulae which form the necessary basis for practical applications.

The second difficulty has been to decide what preliminary mathematics should be included in the volume so that it may be intelligible to those students whose previous mathematical equipment is slight. The general aim of the volumes in the series is that, as far as possible, they shall be self-contained. But in this volume it is obviously necessary to assume *some* previous mathematical training. The study of Trigonometry cannot be begun without a knowledge of Arithmetic, a certain amount of Algebra, and some acquaintance with the fundamentals of Geometry.

It may safely be assumed that all who use this book will have a sufficient knowledge of Arithmetic. In Algebra the student is expected to have studied at least as much as is contained in the volume in this series called *Teach Yourself Algebra*. That work does not include a treatment of " Factors ", but these are not required until Chapter VII. Nor does it touch on quadratic equations; these do not appear however until Chapter XI.

A knowledge of logarithms is, however, indispensable, and there can be no progress in the application of Trigonometry without them. Accordingly Chapter II is devoted to a fairly full treatment of them, and unless the student has studied them previously he should not proceed with the rest of the book until he has mastered this chapter and worked as many of the exercises as possible.

No explanation of graphs has been attempted in this volume. In these days, however, when graphical illustrations enter so generally into our daily life, there can be few who are without some knowledge of them, even if no study has been made of the underlying mathematical principles. But, although graphs of trigonometrical functions are included, they are not essential in general to a working knowledge of the subject.

A certain amount of geometrical knowledge is necessary as a foundation for the study of Trigonometry, and possibly many who use this book will have no previous acquaintance with " Geometry ". For them Chapter I has been included. This chapter is in no sense a course of geometry, or of geometrical reasoning, but merely a brief descriptive account of geometrical terms and of certain fundamental geometrical theorems which will make the succeeding chapters more easily understood. It is not suggested that a great deal of time should be spent on this part of the book, and no exercises are included. It is desirable, however, that the student should make himself well acquainted with the subject-matter of it, so that he is thoroughly conversant with the meaning of the terms employed and acquires something of a working knowledge of the geometrical " theorems " which are stated.

The real study of Trigonometry begins with Chapter III, and from that point until the end of Chapter IX there is very little that can be omitted by any student. Perhaps the only exception is the " Product formulae " in §§ 86–89. This section is necessary, however, for the proof of the important formula of § 98, but a student who is pressed for time and finds this part of the work troublesome, may be content to assume the truth of it when studying § 98. In Chapter IX the student reaches what may be considered the goal of elementary trigonometry, the " solution of the triangle " and its many applications, and there many will be content to stop.

Chapters X, XI and XII are not essential for all practical

INTRODUCTION vii

applications of the subject, but some students, such as electrical engineers and, of course, all who intend to proceed to more advanced work, cannot afford to omit them. It may be noted that previous to Chapter IX only angles which are not greater than $180°$ have been considered, and these have been taken in two stages in Chapters III and V, so that the approach may be easier. Chapter XI continues the work of these two chapters and generalizes with a treatment of angles of any magnitude.

The Exercises throughout have been carefully graded and selected in such a way as to provide the necessary amount of manipulation. Most of them are straightforward and purposeful; examples of academic interest or requiring special skill in manipulation have, generally speaking, been excluded.

Trigonometry employs a comparatively large number of formulae. The more important of these have been collected and printed on pp. 174, 175 in a convenient form for easy reference.

Publisher's Note: This edition has been amended by Harold Frayman to incorporate S.I. units throughout.

CONTENTS

CHAPTER IV

RELATIONS BETWEEN THE TRIGONOMETRICAL RATIOS

CHAPTER V

RATIOS OF ANGLES IN THE SECOND QUADRANT

CHAPTER VI

TRIGONOMETRICAL RATIOS OF COMPOUND ANGLES

CHAPTER VII

RELATIONS BETWEEN THE SIDES AND ANGLES OF A TRIANGLE

CHAPTER VIII

THE SOLUTION OF TRIANGLES

CHAPTER IX

PRACTICAL PROBLEMS INVOLVING THE SOLUTION OF TRIANGLES

CONTENTS

CHAPTER X

CIRCULAR MEASURE

CHAPTER XI

TRIGONOMETRICAL RATIOS OF ANGLES OF ANY MAGNITUDE

CHAPTER XII

TRIGONOMETRICAL EQUATIONS

CHAPTER I

GEOMETRICAL FOUNDATIONS

1. Trigonometry and Geometry.

THE name Trigonometry is derived from the Greek words meaning "triangle" and "to measure". It was so called because in its beginnings it was mainly concerned with the problem of "solving a triangle". By this is meant the problem of finding all the sides and angles of a triangle, when some of these are known.

Before beginning the study of Trigonometry it is very desirable, in order to reach an intelligent understanding of it, to acquire some knowledge of the fundamental geometrical ideas upon which the subject is largely built. Indeed, Geometry itself is thought to have had its origins in practical problems which are now solved by Trigonometry. This is indicated in certain fragments of Egyptian mathematics which are available for our study. We learn from them that from early times Egyptian mathematicians were concerned with the solution of problems arising out of certain geographical phenomena peculiar to that country. Every year the Nile floods destroyed landmarks and boundaries of property. To re-establish them, methods of surveying were developed, and these were dependent upon principles which came to be studied under the name of "Geometry". The word "Geometry", a Greek one, means "Earth measurement", and this serves as an indication of the origins of the subject.

We shall therefore begin by a brief consideration of certain geometrical principles and theorems, the applications of which we shall subsequently employ. It will not be possible, however, within this small book to attempt mathematical proofs of the various theorems which will be stated. The student who has not previously studied the subject of Geometry, and who desires to possess a more complete knowledge of it, should turn to any good modern treatise on this branch of mathematics.

2. The Nature of Geometry.

Geometry has been called "the science of space". It deals with solids, their forms and sizes. By a "*solid*" we mean a "*portion of space bounded by surfaces*", and in Geometry we deal only with what are called "*regular solids*". As a simple example consider that familiar solid, the cube. We are not concerned with the material of which it is composed, but merely the shape of the portion of space which it occupies. We note that it is bounded by six surfaces, which are squares. Each square is said to be at right angles to adjoining squares. Where two squares intersect **straight lines** are formed; three adjoining squares meet in a **point**. These are examples of some of the matters that Geometry considers in connection with this particular solid.

For the purpose of examining the geometrical properties of the solid we employ a conventional representation of the cube, such as is shown in Fig. 1. In this all the faces are shown, as though the body were made of transparent material, those edges which could not otherwise be seen being indicated by dotted lines. The student can follow from this figure the properties mentioned above.

FIG. 1.

3. Plane surfaces.

The surfaces which form the boundaries of the cube are level or flat surfaces, or in geometrical terms "**plane surfaces**". It is important that the student should have a clear idea of what is meant by a plane surface. It may be described as a *level* surface, a term that everybody understands although he may be unable to give a mathematical definition of it. Perhaps the best example in nature of a level surface or plane surface is that of still water. A water surface is also a **horizontal surface**.

The following definition will present no difficulty to the student.

A plane surface is such that the straight line which joins any two points on it lies wholly in the surface.

It should further be noted that

A plane surface is determined uniquely, by

(1) Three points not in the same straight line.
(2) By two intersecting straight lines.

By this we mean that one plane, and one only, can include (1) three given points, or (2) two given intersecting straight lines.

It will be observed that we have spoken of surfaces, points and straight lines without defining them. Every student probably understands what the terms mean, and we shall not consider them further here, but those who would desire more precise knowledge of them should consult a geometrical treatise. We shall proceed to consider theorems connected with points and lines on a plane surface. This is the part of geometry called " *plane geometry* ". The study of the shapes and geometrical properties of solids is the function of " *solid geometry* ", which we will touch on later.

4. Angles are of the utmost importance in Trigonometry, and the student must therefore have a clear understanding of them from the outset. Everybody knows that an angle is formed when two straight lines or two surfaces meet. This has been assumed in § 2. But a precise mathematical definition is desirable. Before proceeding to that, however, we will consider some elementary notions and terms connected with an angle.

In Fig. 2, (*a*), (*b*), (*c*) are shown three examples of angles.

(1) In Fig. 2 (*a*) two straight lines *OA*, *OB*, called the

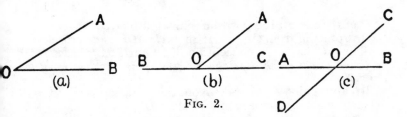

FIG. 2.

arms of the angle, meet at *O* to form the angle denoted by *AOB*.

O is termed the *vertex* of the angle.

The arms may be of any length, and the size of the angle is not altered by increasing or decreasing them.

The " angle *AOB* " can be denoted by $\angle AOB$ or $\widehat{AOB}$. It should be noted that the middle letter, in this case, *O*, always indicates the vertex of the angle.

(2) In Fig. 2 (*b*) the straight line *AO* is said to *meet* the straight line *CB* at *O*. Two angles are formed, *AOB* and *AOC*, with a common vertex *O*.

(3) In Fig. 2 (*c*) two straight lines *AB* and *CD* *cut* one another at *O*. Thus there are formed four angles *COB*, *AOC*, *DOA*, *DOB*.

The pair of angles *COB*, *AOD* are termed *vertically opposite* angles. The angles *AOC*, *BOD* are also vertically opposite.

Adjacent angles. Angles which have a *common vertex* and also *one common arm* are called *adjacent angles*. Thus in Fig. 2 (*b*) *AOB*, *AOC* are adjacent. In Fig. 2 (*c*) *COB*, *BOD* are adjacent, etc.

5. Angles formed by rotation.

We must now consider a mathematical conception of an angle.

Imagine a straight line, starting from a fixed position on OA (Fig. 3), to rotate about a point O in the direction indicated by an arrow.

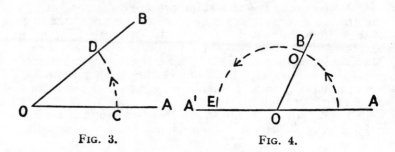

FIG. 3.   FIG. 4.

Let it take up the position indicated by OB.

In rotating from OA to OB an *angle AOB is described.*

Thus we have the conception of an *angle as formed by the rotation of a straight line about a fixed point, the vertex of the angle.*

If any point C be taken on the rotating arm, it will clearly mark out an arc of a circle, CD.

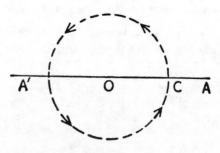

FIG. 5.

There is no limit to the amount of rotation of OA, and consequently *angles of any size* can be formed by a straight line rotating in this way.

A **half rotation.** Let us next suppose that the rotation from OA to OB is continued until the position OA' is reached (Fig. 4), in which OA' and OA are in the same straight line. The point, C, will have marked out a semi-

circle and the angle formed *AOA'* is sometimes called a
" *straight angle* ".

A complete rotation.   Now let the rotating arm con-
tinue to rotate, in the same direction as before, until it
arrives back at its original position on *OA*.   It has then
made a *complete rotation*.   The point *C*, on the rotating arm,
will have marked out the circumference of a circle, as
indicated by the dotted line.

6. Measurement of angles.

(a) Sexagesimal measure.

The conception of formation of an angle by rotation leads
us to a convenient method of measuring angles.   *We*

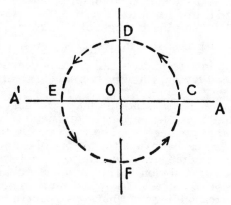

FIG. 6.

imagine the complete rotation to be divided into 360 *equal
divisions*; thus we get 360 small equal angles, each of these
is called a degree, and is denoted by 1°.

Since any point on the rotating arm marks out the circumference of a circle, there will be 360 equal divisions of
this circumference, corresponding to the 360 degrees (see
Theorem 17).   If these divisions are marked on the circumference we could, by joining the points of division to the
centre, show the 360 equal angles.   These could be numbered,
and thus the figure could be used for measuring any given
angle.   In practice the divisions and the angles are very
small, and it would be difficult to draw them accurately.
This, however, is the principle of the " *circular protractor* ",
which is an instrument devised for the purpose of measuring
angles.   Every student of Trigonometry should provide himself with a protractor for this purpose.

Right angles.   Fig. 6 represents a complete rotation, such as was shown in Fig. 5.   Let the points D and F be taken half-way between C and E in each semi-circle.

The circumference is thus divided into four equal parts.

The straight line DF will pass through O.

The angles COD, DOE, EOF, FOC, each one-fourth of a complete rotation, are termed *right angles*, and each contains $90°$.

The circle is divided into four equal parts called *Quadrants*, and numbered the first, second, third and fourth quadrants in the order of their formation.

Also when the rotating line has made a half rotation, the angle formed—the straight angle—must contain $180°$.

Each degree is divided into 60 minutes, shown by $'$.

Each minute is divided into 60 seconds, shown by $''$.

Thus $37° 15' 27''$ means an angle of

$$37 \text{ degrees, } 15 \text{ minutes, } 27 \text{ seconds.}$$

This division into so many small parts is very important in navigation, surveying, gunnery, etc., where great accuracy is essential.

Historical note.   The student may wonder why the number 360 has been chosen for the division of a complete rotation to obtain the degree.   The selection of this number was made in very early days in the history of the world, and we know, for example, from inscriptions that it was employed in ancient Babylon.   The number probably arose from the division of the heavens by ancient astronomers into 360 parts, corresponding to what was reputed to be the number of days in the year.   The number 60 was possibly used as having a large number of factors and so capable of being used for easy fractions.

(*b*) **Centesimal measure.**   When the French adopted the Metric system they abandoned the method of dividing the circle into 360 parts.   To make the system of measuring angles consistent with other metric measures, it was decided to divide the right angle into 100 equal parts, and consequently the whole circle into 400 parts.   The angles thus obtained were called grades.

Consequently 1 right angle = 100 grades.
          1 grade   = 100 minutes.
          1 minute   = 100 seconds.

(*c*) **Circular measure.**   There is a third method of measuring angles which is an *absolute* one, that is, it does not depend upon dividing the right angle into any arbitrary number of equal parts, such as 360 or 400.

The unit is obtained as follows:

In a circle, centre O (see Fig. 7), let a radius OA rotate to a position OB, such that,

the length of the *arc* AB is equal to that of the radius.

In doing this an angle AOB is formed which is the unit of measurement. It is called a *radian*. The size of this angle will be the same whatever radius is taken. It is absolute in magnitude.

In degrees 1 *radian* = *57° 17′ 44·8″* (approx.). This method of measuring angles will be dealt with more fully in Chapter X. It is very important and is always used in the higher branches of mathematics.

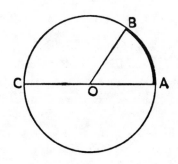

Fig. 7.

7. Terms used to describe angles.

An **Acute angle** is an angle which is **less than a right angle**.

An **Obtuse angle** is one which is **greater than a right angle**.

Reflex or **re-entrant angles** are angles between 180° and 360°.

Complementary angles. When the sum of two angles is equal to a right angle, each is called the **complement** of the other. Thus the complement of 38° is 90° − 38° = 52°.

Supplementary angles. When the sum of two angles is equal to 180°, each angle is called the **supplement** of the other. Thus the supplement of 38° is 180° − 38° = 142°.

8. Geometrical Theorems.

We will now proceed to state, without proof, some of the more important geometrical theorems.

Theorem I. Intersecting straight lines.

If two straight lines intersect, the vertically opposite angles are equal. (See § 4.)

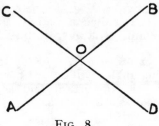

FIG. 8.

In Fig. 8, AB and CD are two straight lines intersecting at O.

Then $$\angle AOC = \angle BOD$$
and $$\angle COB = \angle AOD.$$

The student will probably see the truth of this on noticing that $\angle AOC$ and $\angle BOD$ are each supplementary to the same angle, COB.

9. Parallel straight lines.

Take a set square PRQ (Fig. 9) and slide it along the edge of a ruler.

Let $P_1R_1Q_1$ be a second position which it takes up.

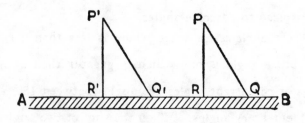

FIG. 9.

It is evident that the inclination of PQ to AB is the same as that of P_1Q_1 to AB; since there has been no change in direction.

$$\therefore \quad \angle PQB = \angle P_1Q_1B$$

If PQ and P_1Q_1 were produced to any distance they would not meet.

The straight lines PQ and P_1Q_1 are said to be parallel.

Similarly PR and P_1R_1 are parallel.

Hence the following definition.

Straight lines in the same plane which will not meet however far they may be produced are said to be parallel.

Direction. *Parallel straight lines in a plane have the same direction.* If a number of ships, all sailing North in a convoy are ordered to change direction by turning through the same angle, they will then follow parallel courses.

Terms connected with parallel lines.

In Fig. 10 AB, CD represent two parallel straight lines.

Transversal. A straight line such as PQ which cuts them is called a transversal.

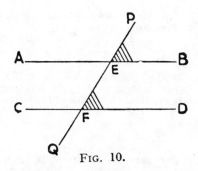

Fig. 10.

Corresponding angles. On each side of the transversal are two pairs of angles, one pair of which is shaded in the figure. These are called corresponding angles.

Alternate angles. Two angles such as AEF, EFD on opposite sides of the transversal, are called alternate angles.

Theorem 2.

If a pair of parallel straight lines be cut by a transversal

(1) Alternate angles are equal.

(2) Corresponding angles on the same side of the transversal are equal.

(3) The two interior angles on the same side of the transversal are equal to two right angles.

Thus in Fig. 10.

Alternate angles. $\angle AEF = \angle EFD$; $\angle BEF = \angle EFC$.

Corresponding angles.

$\angle PEB = \angle EFD$; $\angle BEF = \angle DFQ$.

Similarly on the other side of the transversal.

Interior angles $\angle BEF + \angle EFD = 2$ right angles.

also $\qquad \angle AEF + \angle EFC = 2$ right angles.

10. Triangles.

Kinds of triangles.

 A *right-angled triangle* has one of its angles a right angle. The side opposite to the right angle is called the *hypotenuse*.

 An *acute-angled triangle* has all its angles acute angles (see § 7).

 An *obtuse-angled triangle* has one of its angles obtuse (see § 7).

 An *isosceles triangle* has two of its sides equal.

 An *equilateral triangle* has all its sides equal.

FIG. 11.

Lines connected with a triangle. The following terms are used for certain lines connected with a triangle.

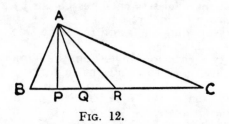

FIG. 12.

In △ ABC, Fig. 12,

(1) *AP* is the perpendicular from *A* to *BC*. It is called the *altitude* from the vertex *A*.

(2) *AQ* is the *bisector of the vertical angle* at *A*.

(3) *AR* bisects *BC*. It is called a *median*. If each of the points *B* and *C* be taken as a vertex, there are two other corresponding medians. Thus a triangle may have three medians.

11. Theorem 3.   Isosceles and equilateral triangles.

In an isosceles triangle

(*a*) The sides opposite to the equal angles are equal.

(*b*) A straight line drawn from the vertex perpendicular to the opposite side bisects that side and the vertical angle.

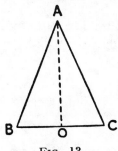

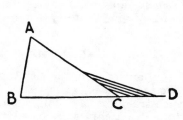

| FIG. 13. | FIG. 14. |

In Fig. 13, *ABC* is an isosceles △ and *AO* is drawn perp. to the base from the vertex *A*.

Then by the above $\angle ABC = \angle ACB$
$$BO = OC$$
$$\angle BAO = \angle CAO.$$

Equilateral triangle. The above is true for an equilateral triangle, and since all its sides are equal, all its angles are equal.

Note.—In an isosceles △ the altitude, median and bisector of the vertical angle (see § 10) coincide when the point of intersection of the two equal sides is the vertex. If the △ is equilateral they coincide for all three vertices.

12. Angle properties of a triangle.

Theorem 4.   If one side of a triangle be produced, the exterior angle so formed is equal to the sum of the two interior opposite angles.

Thus in Fig. 14 one side *BC* of the △ *ABC* is produced to *D*.

$\angle ACD$ is called an exterior angle.

Then by the above

$$\angle ACD = \angle ABC + \angle BAC$$

Notes.—(1) Since the exterior angle is equal to the sum of the opposite interior angles, it must be greater than either of them.

(2) As each side of the triangle may be produced in turn, there are three exterior angles.

Theorem 5. The sum of the angles of any triangle is equal to two right angles.

Notes.—It follows:

(1) Each of the angles of an equilateral triangle is 60°.

(2) In a right-angled triangle the two acute angles are complementary (see § 7).

(3) The sum of the angles of a quadrilateral is 360°, since it can be divided into two triangles by joining two opposite points.

13. Congruency of triangles.

Definition. Triangles which are equal in all respects are said to be congruent.

Such triangles have corresponding sides and angles equal, and are exact copies of one another.

If two triangles ABC and DEF are congruent we may express this by the notation $\triangle ABC \equiv \triangle DEF$.

Conditions of congruency.

Two triangles are congruent when

(1) **Theorem 6. Three sides of one are respectively equal to the three sides of the other.**

(2) **Theorem 7. Two sides of one and the angle they contain are equal to two sides and the contained angle of the other.**

(3) **Theorem 8. Two angles and a side of one are equal to two angles and the corresponding side of the other.**

These conditions in which triangles are congruent are very important. The student can test the truth of them practically by constructing triangles which fulfil the conditions stated above.

The ambiguous case.

The case of constructing a triangle when there are *given two sides and an angle opposite to one of them*, not contained by them as in Theorem 7, requires special consideration.

Example. Construct a triangle in which two sides are

35 mm and 25 mm and the angle opposite the smaller of these is 30°.

The construction is as follows:

Draw a straight line AX of indefinite length (Fig. 15).

At A construct $\angle BAX = 30°$ and make $AB = 35$ mm.

With B as centre and radius 25 mm construct an arc of a circle to cut AX.

This it will do in two points, C and C'.

Consequently if we join BC or BC' we shall complete two triangles ABC, ABC' each of which will fulfil the given conditions. There being thus two solutions the case is called " ambiguous ".

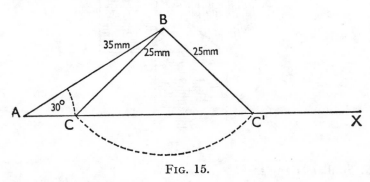

FIG. 15.

14. Right-angled triangles.

Theorem of Pythagoras.

Theorem 9. In every right-angled triangle the square on the hypotenuse is equal to the sum of the squares on the sides containing the right angle.

In Fig. 16 ABC is a right-angled triangle, AB being the hypotenuse. On the three sides squares have been constructed. Then the area of the square described on AB is equal to the sum of the areas of the squares on the other two sides.

This we can write in the form

$$AB^2 = AC^2 + BC^2.$$

If we represent the length of AB by c, AC by b and BC by a, then $c^2 = a^2 + b^2$.

It should be noted that by using this result, if any two sides of a right-angled triangle are known, we can find the other side, for

$$a^2 = c^2 - b^2$$
$$b^2 = c^2 - a^2.$$

Note.—This theorem is named after *Pythagoras*, the Greek mathematician and philosopher who was born about 569 B.C. It is one of the most important and most used of all geometrical theorems. Two proofs are given in Vol. I of *National Certificate Mathematics*, published by the English Universities Press.

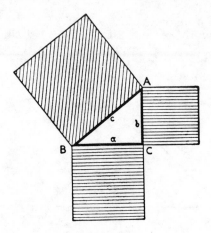

FIG. 16.

15. Similar triangles.

Definition. If the angles of one triangle are respectively equal to the angles of another triangle the two triangles are said to be similar.

The sides of similar triangles which are opposite to equal angles in each are called **corresponding sides**.

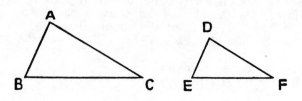

FIG. 17.

In Fig. 17 the triangles *ABC*, *DEF* are equiangular
$$\angle ABC = \angle DEF,$$
$$\angle BAC = \angle EDF,$$
$$\angle ACB = \angle DFE.$$

The sides AB, DE are two corresponding sides.
So also are AC and DF, BC and EF.
Fig. 18 shows another example of interest later.
AB, CD, EF are parallel.
Then by the properties of parallel lines (see § 9)

$$\angle OAB = \angle OCD = \angle OEF$$
also
$$\angle OBA = \angle ODC = \angle OFE.$$

∴ the triangles OAB, OCD, OEF are similar.

Property of similar triangles.

Theorem 10. If two triangles are similar, the corresponding sides are proportional.

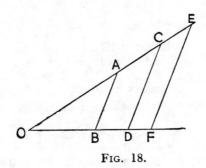

FIG. 18.

Thus in Fig. 17 :

$$\frac{AB}{BC} = \frac{DE}{EF}, \quad \frac{AB}{AC} = \frac{DE}{DF}, \quad \frac{AC}{CB} = \frac{DF}{FE}$$

Similarly in Fig. 18 :

$$\frac{AB}{BO} = \frac{CD}{DO} = \frac{EF}{FO},$$

$$\frac{AB}{OA} = \frac{CD}{OC} = \frac{EF}{OE}, \text{ etc.}$$

These results are of great importance in Trigonometry.

Note.—A similar relation holds between the sides of quadrilaterals and other rectilinear figures which are equiangular.

16. Quadrilaterals.

A *quadrilateral* is a plane figure with four sides, and a straight line joining two opposite angles is called a *diagonal*.

The following are among the principal quadrilaterals, with some of their properties :

(1) The **square** (*a*) has all its sides equal and all its angles right angles; (*b*) its diagonals are equal, bisect each other at right angles and also bisect the opposite angles.

(1)

(2) The **rhombus** (*a*) has all its sides equal; (*b*) its angles are not right angles; (*c*) its diagonals bisect each other at right angles and bisect the opposite angles.

(2)

(3) The **rectangle** (*a*) has opposite sides equal and all its angles are right angles; (*b*) its diagonals are equal and bisect each other.

(3)

(4) The **parallelogram** (*a*) has opposite sides equal and parallel; (*b*) its opposite angles are equal; (*c*) its diagonals bisect each other.

(4)

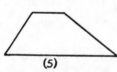

(5) The **trapezium** has two opposite sides parallel.

(5)

Fig. 19.

17. The Circle.

Definitions. It has already been assumed that the student understands what a circle is, but we now give a geometrical definition.

A **circle** *is a plane figure bounded by one line which is called the circumference and is such that all straight lines drawn to the circumference from a point within the circle, called the centre, are equal.*

These straight lines are called *radii*.

An **arc** is a part of the circumference.

A **chord** is a straight line joining two points on the circumference and dividing the circle into two parts.

A **diameter** is a chord which passes through the centre of the circle. It divides the circle into two equal parts called *semi-circles*.

A **segment** is a part of a circle bounded by a chord and the arc which it cuts off. Thus in Fig. 20 the chord *PQ* divides the circle into two segments. The larger of these *PCQ* is called a *major segment* and the smaller, *PBQ*, is called a *minor segment*.

A **sector** of a circle is that part of the circle which is bounded by two radii and the arc intercepted between them.

Thus in Fig. 21 the figure *OPBQ* is a sector bounded by the radii *OP*, *OQ* and the arc *PBQ*.

An **angle in a segment** is the angle formed by joining the ends of a chord or arc to a point on the arc of the segment.

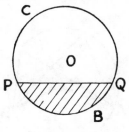
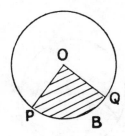

FIG. 20.   FIG. 21.

Thus in Fig. 22, the ends of the chord *AB* are joined to *D* a point on the arc of the segment. The angle *ADB* is the angle in the segment *ABCD*.

If we join *A* and *B* to any point *D'* in the minor segment, then ∠*AD'B* is the angle in the minor segment.

If *A* and *B* are joined to the centre *O*, the angle *AOB* is called the *angle at the centre*.

The angle *ADB* is also said to *subtend* the arc *AB* and the ∠*AOB* is said to be the angle *subtended* at the centre by the arc *AB* or the chord *AB*.

Concentric Circles. Circles which have the same centre are called concentric circles.

18. Theorems relating to the circle.

Theorem 11. If a diameter bisects a chord, which is not a diameter, it is perpendicular to the chord.

Theorem 12. Equal chords in a circle are equidistant from the centre.

Theorem 13. The angle which is subtended at the

centre of a circle by an arc is double the angle subtended at the circumference.

In Fig. 23 $\angle AOB$ is the angle subtended at O the centre of the circle by the arc AB, and $\angle ADB$ is an angle at the circumference (see § 17) as also is $\angle ACB$.

Then $\angle AOB = 2\angle ADB$.
Also $\angle AOB = 2\angle ACB$.

Theorem 14. Angles in the same segment of a circle are equal to one another.

In Fig. 23 $\angle ACB = \angle ADB$.

This follows at once from Theorem 13.

Theorem 15. The opposite angles of a quadrilateral inscribed in a circle are together equal to two right angles.

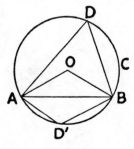

Fig. 22.

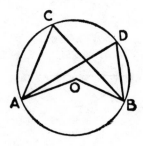

Fig. 23.

They are therefore supplementary (see § 7).

Note.—A quadrilateral inscribed in a circle is called a cyclic or concyclic quadrilateral.

In Fig. 24, $ABCD$ is a cyclic quadrilateral.

Then $\angle ABC + \angle ADC = 2$ right angles
            $\angle BAD + \angle BCD = 2$ right angles.

Theorem 16. The angle in a semi-circle is a right angle.

In Fig. 25 AOB is a diameter.

The $\angle ACB$ is an angle in one of the semi-circles so formed. $\angle ACB$ is a right angle.

Theorem 17. Angles at the centre of a circle are proportional to the arcs on which they stand.

In Fig. 26,

$$\frac{\angle POQ}{\angle QOR} = \frac{\text{arc } PQ}{\text{arc } QR}.$$

It follows from this that *equal angles stand on equal arcs*.

This is assumed in the method of measuring angles described in § 6(*a*).

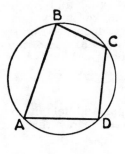

FIG. 24.

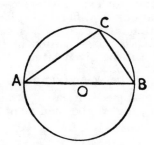

FIG. 25.

Tangent to a circle.

A tangent to a circle is a straight line which meets the circumference of the circle but which when produced does not cut it.

In Fig. 27 *PQ* represents a tangent to the circle at a point *A* on the circumference.

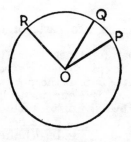

FIG. 26.

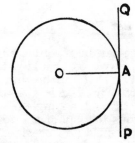

FIG. 27.

Theorem 18. A tangent to a circle is perpendicular to the radius drawn from the point of contact.

Thus in Fig. 27 *PQ* is at right angles to *OA*.

SOLID GEOMETRY

19. We have so far confined ourselves to the consideration of some of the properties of figures drawn on plane surfaces. In many of the practical applications of Geometry we are concerned also with " *solids* " to which we have referred in § 2. In addition to these, in surveying and navigation problems, for example, we need to make observations and calculations in different " *planes* ", which are not specifically the surfaces of solids. Examples of these, together with a brief classification of the different kinds of regular solids, will be given later.

20. Angle between two planes.

Take a piece of fairly stout paper and fold it in two. Let *AB*, Fig. 28, be the line of the fold. Draw this straight

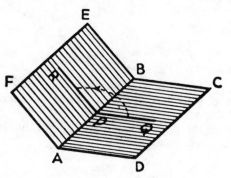

Fig. 28.

line. Let *BCDA*, *BEFA* represent the two parts of the paper.

These can be regarded as two separate planes. Starting with the two parts folded together, keeping one part fixed the other part can be rotated about *AB* into the position indicated by *ABEF*. In this process the one plane has moved through an angle relative to the fixed plane. This is analogous to that of the rotation of a line as described in § 5. We must now consider how this angle can be definitely fixed and measured. Flattening out the whole paper again take any point *P* on the line of the fold, *i.e. AB*, and draw *RPQ* at right angles to *AB*. If you fold again *PR* will coincide with *PQ*. Now rotate again and the line *PR* will mark out an angle relative to *PQ* as we saw in § 5. The angle *RPQ* is thus the angle which measures the amount of rotation, and is called the angle between the planes.

Definition. The angle between two planes is the angle between two straight lines which are drawn, one in each plane, at right angles to the line of intersection of the plane and from the same point on it.

When this angle becomes a right angle the *planes are perpendicular to one another.*

As a particular case a plane which is perpendicular to a horizontal plane is called a *vertical plane* (see § 3).

If you examine a corner of the cube shown in Fig. 1 you will see that it is formed by three planes at right angles to one another. A similar instance may be observed in the corner of a room which is rectangular in shape.

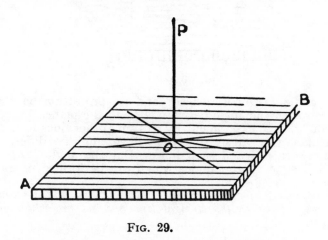

Fig. 29.

21. A straight line perpendicular to a plane.

Take a piece of cardboard *AB* (Fig. 29), and on it draw a number of straight lines intersecting at a point *O*. At *O* fix a pin *OP* so that it is perpendicular to all of these lines. Then *OP* is said to be perpendicular to the plane *AB*.

Definition. A straight line is said to be perpendicular to a plane when it is perpendicular to any straight line which it meets in the plane.

Plumb line and vertical. Builders use what is called a plumb line to obtain a vertical line. It consists of a small weight fixed to a fine line. This vertical line is perpendicular to a horizontal plane.

22. Angle between a straight line and a plane.

Take a piece of cardboard *ABCD*, Fig. 30, and at a point *O* in it fix a needle *ON* at any angle. At any point *P* on the needle stick another needle *PQ* into the board, and perpendicular to the board.

Draw the line *OQR* on the board.

OQ is called the projection of OP on the plane ABCD.

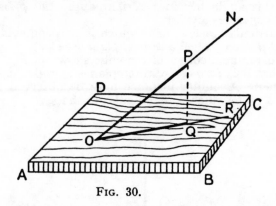

FIG. 30.

The angle *POQ* between *OP* and its projection on the plane is called the angle between *OP* and the plane.

If you were to experiment by drawing other lines from *O* on the plane you will see that you will get angles of different sizes between *ON* and such lines. But the angle *POQ* is the smallest of all the angles which can be formed in this way.

Definition. The angle between a straight line and a plane is the angle between the straight line and its projection on the plane.

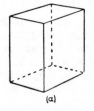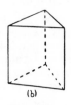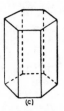

(a)   (b)   (c)

FIG. 31.

23. Some regular solids.

(1) **Prisms.** In Fig. 31(*a*), (*b*), (*c*) are shown three typical prisms.

(*a*) is rectangular, (*b*) is triangular and (*c*) is hexagonal.

They have two ends or bases, identically equal and a rectangle, triangle and regular hexagon respectively.

The sides are rectangles in all three figures and their planes are perpendicular to the bases.

Such prisms are called *right prisms*.

If sections are made parallel to the bases, all such sections are identically equal to the bases. A prism is a solid with a *uniform cross section*.

Similarly other prisms can be constructed with other geometrical figures as bases.

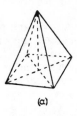 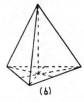 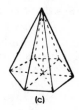

(a) (b) (c)

FIG. 32.

(2) **Pyramids.** In Fig. 32 (*a*), (*b*), (*c*), are shown three typical pyramids.

(*a*) is a square pyramid;

(*b*) is a triangular pyramid;

(*c*) is a hexagonal pyramid.

Pyramids have one base only, which, as was the case with prisms, is some geometrical figure.

The sides, however, are isosceles triangles, and they meet at a point called the **vertex**.

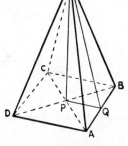

FIG. 33.

The *angle between each side and the base* can be determined as follows for a square pyramid.

In Fig. 33, let P be the intersection of the diagonals of the base.

Join P to the vertex O.

When OP is perpendicular to the base the pyramid is a *right* pyramid and OP is its *axis*.

Let Q be the mid-point of one of the sides of the base AB.

Join PQ and OQ.

Then PQ and OQ are perpendicular to AB (Theorem 11).

It will be noticed that OPQ represents a plane, imagined within the pyramid but not necessarily the surface of a solid.

Then by the Definition in § 20, the angle OQP represents the angle between the plane of the base and the plane of the side OAB.

Clearly the angles between the other sides and the base will be equal to this angle.

Note.—This angle must not be confused with angle *OBP* which students sometimes take to be the angle between a side and the base.

Sections of right pyramids.

If sections are made parallel to the base, and therefore at right angles to the axis, they are of the same shape as the base, but of course smaller and similar.

(3) Solids with curved surfaces.

The surfaces of all the solids considered above are plane surfaces. There are many solids whose surfaces are either entirely curved or partly plane and partly curved. Three well-known ones can be mentioned here, the cylinder, the cone and the sphere. Sketches of two of these are shown below in Fig. 34(a) and (b).

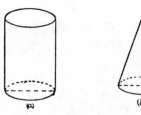

FIG. 34.

(a) The **cylinder** (Fig. 34(a)). This has two bases which are equal circles and a curved surface at right angles to these. A cylinder can be easily made by taking a rectangular piece of paper and rolling it round until two ends meet. This is sometimes called a circular prism.

(b) The **cone** (Fig. 34(b)). This is in reality a pyramid with a circular base.

(c) The **sphere**. A sphere is a solid such that any point on its surface is the same distance from a point within, called the centre. Any section of a sphere is a circle.

24. Angles of elevation and depression.

The following terms are used in practical applications of Geometry and Trigonometry.

(a) Angle of elevation.

Suppose that a surveyor, standing at *O* (Fig. 35) wishes to determine the height of a distant tower and spire. His first step would be to place a telescope (in a theodolite) horizontally at *O*. He would then rotate it *in a vertical plane* until it pointed to the top of the spire. The angle through which he rotates it, the angle *POQ*, in Fig. **35** is called the *angle of elevation* or the *altitude* of *P*.

Sometimes this is said to be the angle *subtended* by the building at *O*.

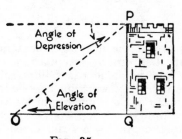

FIG. 35.

Altitude of the sun. The *altitude* of the sun is in reality the angle of elevation of the sun. It is the angle made by the sun's rays, considered parallel, with the horizontal at any given spot at a given time.

(*b*) **Angle of depression.**

If at the top of the tower shown in Fig. **35**, a telescope were to be rotated from the horizontal till it points to an object at *O*, the angle so formed is called the *angle of depression*.

CHAPTER II

LOGARITHMS

25. LOGARITHMS are of the utmost importance in Trigonometry. Without them many calculations would be extremely tedious and in some cases impossible. Lest the student should not have a working knowledge of them we give a brief summary of their nature, properties and uses.

Logarithms are Indices viewed from a special standpoint. We must therefore begin by a brief consideration of the laws of Indices.

It will be learnt from *Teach Yourself Mathematics* that a^4 represents $a \times a \times a \times a$, where a is any number.

The Index " 4 " indicates the number of factors.

Generally, if " n " stands for any whole number

a^n means $a \times a \times a \times \ldots$ to n factors and
a^n is called the nth power of a.

26. Laws of Indices.

We now proceed to the laws which govern the use of Indices.

(1) **Law of multiplication.**

Since $\quad a^4 = a \times a \times a \times a$ (*i.e.* the product of " 4 a's ")
and $\quad\quad a^3 = a \times a \times a$ (the product of " 3 a's ").
then $\quad\quad a^4 \times a^3 =$ the product of $(4 + 3)a$'s.
i.e. $\quad\quad\quad\quad a^4 \times a^3 = a^{4+3}$
$\quad\quad\quad\quad\quad\quad\quad\quad = a^7.$

And generally if m and n are any positive integers we can prove

$$a^m \times a^n = a^{m+n}.$$

This law is obviously true for any number of factors, *e.g.*
$$a^m \times a^n \times a^p = a^{m+n+p}.$$

(2) **Law of division.**

Since $\quad\quad\quad a^5 = a \times a \times a \times a \times a$
and $\quad\quad\quad a^3 = a \times a \times a$

on division the three factors of a^3 cancel three of the five factors of a^5.

Thus $(5 - 3)$ *i.e.* 2 factors are left.

$$\therefore \qquad a^5 \div a^3 = a^{5-3}$$

and in general we can prove

$$a^m \div a^n = a^{m-n}.$$

(3) Law of powers.

Suppose we require the value of $(a^5)^3$, *i.e.* the third power of a^5. This by the meaning of an index is

$$a^5 \times a^5 \times a^5$$

and by the first law of Indices, above,

$$a^5 \times a^5 \times a^5 = a^{5+5+5}$$

i.e.
$$(a^5)^3 = a^{5\times 3}$$
$$= a^{15}.$$

In general $\qquad (a^m)^n = a^{mn}.$

27. Summary of the Laws of Indices.

(1) Multiplication. $\qquad a^m \times a^n = a^{m+n}.$
(2) Division. $\qquad a^m \div a^n = a^{m-n}.$
(3) Powers. $\qquad (a^m)^n = a^{mn}.$

Exercise 1.

1. Write down the values of :

    (1) $a^4 \times a^6.$ (4) $\frac{1}{2}x \times \frac{1}{4}x^7 \times \frac{3}{4}x^2.$
    (2) $b^7 \times b^5.$ (5) $2^3 \times 2^4.$
    (3) $x^3 \times x^4 \times x^5.$ (6) $3 \times 3^2 \times 3^4.$

2. Write down the values of :

    (1) $a^7 \div a^3.$ (3) $x^{16} \div x^4.$
    (2) $c^{10} \div c^5.$ (4) $2^{10} \div 2^4.$

3. Find the values of:

    (1) $x^7 \times x^4 \div x^5.$ (3) $\dfrac{a^7}{a^5} \times \dfrac{a}{a^4}.$

    (2) $a^6 \times a^5 \div a^8.$ (4) $\dfrac{x^6 \times x^4}{x^3}.$

4. Find the values of :

    (1) $(a^7)^2.$ (5) $(10^2)^3.$
    (2) $(x^4)^3.$ (6) $(3a^2)^3.$
    (3) $(2b^4)^4.$ (7) $(\frac{1}{5}x^4)^5.$
    (4) $(2^4)^2.$ (8) $(3^3)^3.$

28. Extension of the meaning of an index.

The student will readily understand how useful and important indices are in Algebra. He will note that so far they have been restricted to positive whole numbers only, and the meaning given to such a quantity as a^n is unintelligible except on the supposition that n is a positive integer. But we will now consider the possibility of extending the uses of indices so that they can have any value.

The student may already have noticed one instance which will be among those we shall consider in detail later. If we divide a^3 by a^5 and write this down in the form $\dfrac{a \times a \times a}{a \times a \times a \times a \times a}$, we obtain on cancelling, $\dfrac{1}{a \times a}$ or $\dfrac{1}{a^2}$.

If a^3 be divided by a^5 according to rule we have

$$a^3 \div a^5 = a^{3-5}$$
$$= a^{-2}$$

We are thus left with a negative index. But the working above shows that the result of the division of a^3 by a^5 is $\dfrac{1}{a^2}$.

Consequently it appears that a^{-2} means the same thing as $\dfrac{1}{a^2}$, or the reciprocal of a^2.

Thus it seems that a meaning can be given to a^{-2} which is, of course, quite different from the meaning when the index is a positive whole number. We are therefore led to consider what meanings can be given in all those cases in which the index is not a positive integer. In seeking these meanings of an index there is one fundamental principle which will always guide us, viz. : *Every index must obey the laws of indices as discovered for positive integers*. In other words, we will assume that the laws of indices, as stated above, are true in all cases.

29. Fractional Indices.

We will begin with the simple case of $a^{\frac{1}{2}}$. Since, by the above principle, it must conform to the laws of Indices, then, applying the law of multiplication

$$a^{\frac{1}{2}} \times a^{\frac{1}{2}} = a^{\frac{1}{2}+\frac{1}{2}}$$
$$= a^1 \text{ or } a$$

∴ $a^{\frac{1}{2}}$ must be such a quantity that, on being multiplied by itself, the result is a.

∴ $a^{\frac{1}{2}}$ *must be defined as the square root of a*

$$\text{or } a^{\frac{1}{2}} = \sqrt{a}$$

Similarly

$$a^{\frac{1}{3}} \times a^{\frac{1}{3}} \times a^{\frac{1}{3}} = a^{\frac{1}{3}+\frac{1}{3}+\frac{1}{3}}$$
$$= a$$

(First law of indices)

$\therefore$ $a^{\frac{1}{3}}$ *must be defined as the cube root of a.*

The same argument may be applied in other cases, and so generally

$$a^{\frac{1}{n}} = \sqrt[n]{a}$$

To find a meaning for $a^{\frac{2}{3}}$

Applying the first law of indices

$$a^{\frac{2}{3}} \times a^{\frac{2}{3}} \times a^{\frac{2}{3}} = a^{\frac{2}{3}+\frac{2}{3}+\frac{2}{3}}$$
$$= a^2$$

$\therefore$ $a^{\frac{2}{3}}$ *must be the cube root of a^2*

or $\qquad\qquad a^{\frac{2}{3}} = \sqrt[3]{a^2}$

Similarly $\qquad\quad a^{\frac{3}{4}} = \sqrt[4]{a^3}$

and generally $\qquad a^{\frac{m}{n}} = \sqrt[n]{a^m}$

The student will note that decimal indices can be reduced to vulgar fractions and defined accordingly.

Thus $\qquad\qquad a^{0.25} = a^{\frac{1}{4}}$
$$= \sqrt[4]{a}$$

30. To find a meaning for a^0

$$a^n \div a^n = 1$$

But, using the law of division for Indices,

$$a^n \div a^n = a^{n-n}$$
$$= a^0$$
$$\therefore \quad a^0 = 1$$

It should be noted that a represents any number. This result therefore is independent of the value of a.

31. Negative indices.

To find a meaning for a^{-n}
$$a^{-n} \times a^n = a^{-n+n} \qquad\text{(First law of indices)}$$
$$= a^0$$
$$= 1 \text{ (shown above)}$$

Dividing by a^n

$$a^{-n} = \frac{1}{a^n}$$

We may therefore define a^{-n} as the reciprocal of a^n.
Examples

$$a^{-1} = \frac{1}{a}$$

$$2a^{-3} = \frac{2}{a^3}$$

$$a^{-\frac{1}{2}} = \frac{1}{\sqrt{a}}$$

$$\frac{1}{a^{-2}} = a^2$$

or generally $\quad \dfrac{1}{a^{-n}} = a^n.$

Exercise 2

Where necessary in the following take $\sqrt{2} = 1\cdot414$, $\sqrt{3} = 1\cdot732$, $\sqrt{10} = 3\cdot162$, each correct to three places of decimals.

1. Write down the meanings of :

$$3^{\frac{1}{2}}, \ 4^{-1}, \ 3a^{-2}, \ 1000^0, \ 2^{-\frac{1}{2}}, \ \frac{1}{3^{-1}}, \ \frac{3}{a^{-2}}, \ 4^{\frac{3}{2}}, \ 10^{-3}.$$

2. Find the values of :
   (1) $2^2 \times 2^{\frac{1}{2}}$.
   (2) $3 \times 3^{\frac{1}{2}} \times 3^{\frac{3}{2}}$.
   (3) $10^{\frac{1}{2}} \div 10^{\frac{3}{2}}$.
   (4) $a^{\frac{1}{2}} \times a^{\frac{3}{2}}$.
   (5) $2^{\frac{3}{2}}$.
   (6) $10^{\frac{1}{2}}$.

3. Find the values of :
   (1) $8^{\frac{1}{3}}$.
   (2) $25^{\frac{3}{2}}$.
   (3) $(10^2)^{\frac{3}{2}}$.
   (4) $(5^{-3})^2$.
   (5) $\dfrac{2}{2^{-3}}$.
   (6) $(1000)^{\frac{1}{3}}$.

4. Find the values of :
   (1) $(\frac{1}{2})^{-2}$.
   (2) $(\frac{2}{3})^{-3}$.
   (3) $(16)^{0\cdot5}$.
   (4) $(36)^{-0\cdot5}$.
   (5) $(4)^{1\cdot5}$.
   (6) $(\frac{1}{4})^{2\cdot5}$.

5. Find the value of $a^4 \times a^{-2} \times a^{\frac{1}{2}}$ when $a = 2$.

6. Write down the simplest form of :
   (1) $a^{\frac{1}{2}} \times a^{\frac{1}{2}}$.
   (2) $10^3 \times 10^{-\frac{1}{2}}$.

32. A system of logarithms.

These extensions of the meanings of indices to all kinds of numbers are of great practical importance. They enable us to carry out, easily and accurately, calculations which without them would be almost impossible or very laborious. We will choose a very simple example to explain how they can be used.

We have seen that

$$10^{0.5} = 10^{\frac{1}{2}} = \sqrt{10} = 3.162 \text{ approx.} \quad \text{(by calculation)}$$

Now $\quad 10^{1.5} = 10^{1+\frac{1}{2}} = 10^1 \times 10^{\frac{1}{2}}$ (First law of Indices)
$$= 10 \times 3.162 = 31.62$$

Also $\quad 10^{\frac{1}{4}} = (10^{\frac{1}{2}})^{\frac{1}{2}}$ (Third law of Indices)
$$= \sqrt{10^{\frac{1}{2}}} = \sqrt{3.162}$$
$$= 1.78 \text{ approx.}$$

Again $\quad 10^{\frac{3}{4}} = 10^{\frac{1}{2}+\frac{1}{4}} = 10^{\frac{1}{2}} \times 10^{\frac{1}{4}}$ (First law of Indices)
$$= 3.162 \times 1.78$$
$$= 5.62 \text{ approx.}$$

Also $\quad 10^{\frac{1}{8}} = (10^{\frac{1}{4}})^{\frac{1}{2}} = \sqrt{10^{\frac{1}{4}}}$
$$= \sqrt{1.78} = 1.33 \text{ approx.}$$

Similarly we might calculate a number of powers of 10. Let us now make a table showing (1) the numbers above, (2) the *indices showing what powers they are of 10.*

Number.	Index.
1·33	0·125
1·78	0·25
3·162	0·5
5·62	0·75
31·62	1·5

Now suppose we want to find the value of
$$3.162 \times 1.78$$

From the table we see that $3.162 = 10^{0.5}$
$$1.78 = 10^{0.25}$$
$$\therefore \quad 3.162 \times 1.78 = 10^{0.5} \times 10^{0.25}$$
$$= 10^{0.5+0.25} \quad \text{(First law of Indices)}$$
$$= 10^{0.75}$$

Now the table shows us that $10^{0.75} = 5.62$
$$\therefore \quad 3.162 \times 1.78 = 5.62$$

(*Note.*—All the numbers calculated are approximate.)

You will see *instead of the process of multiplication of the numbers, we use that of an addition of the indices.*

Much more difficult calculations can be similarly performed.

It is evident that if we are to make an extended use of this method, one thing is essential.

We must have a table of the indices which indicate the power any given number is of a selected number such as 10 which we have used for the above example.

Such a table is called a *table of logarithms* and the number, such as 10 used above, with respect to which the logarithms are calculated, is called the *base of the system*.

We can therefore define a logarithm as follows.

Definition. A logarithm of a number to a given base is the index of the power to which the base must be raised to produce the number.

For example, we know that

$$341 = 10^{2 \cdot 5328} \tag{1}$$

Then by the above definition

2·5328 *is the logarithm of* 341 *to the base* 10

This we abbreviate into

$$2 \cdot 5328 = \log_{10} 341 \tag{2}$$

the base 10 being indicated by the suffix, as shown. The student should carefully note that equations (1) and (2) are two ways of expressing the same relation between the numbers employed.

33. Characteristic of a logarithm.

The integral or whole number part of a logarithm is called the characteristic. This can always be determined by inspection when logarithms are calculated to base 10, as will be seen from the following considerations:

Since
$$10^0 = 1, \qquad \log_{10} 1 \qquad = 0$$
$$10^1 = 10, \qquad \log_{10} 10 \qquad = 1$$
$$10^2 = 100, \qquad \log_{10} 100 \qquad = 2$$
$$10^3 = 1000, \qquad \log_{10} 1000 \qquad = 3$$
$$10^4 = 10,000, \quad \log_{10} 10,000 = 4$$

and so on.

From these results we see that,

for numbers between   1 and   10 the characteristic is 0
,,   ,,   ,,   10 ,,   100 ,,   ,,   ,, 1
,,   ,,   ,,   100 ,,   1000 ,,   ,,   ,, 2
,,   ,,   ,,   1000 ,,   10,000 ,,   ,,   ,, 3

and so on.

It is evident that the characteristic is always one less than the number of digits in the whole number part of the number.

Thus in $\log_{10}$ 3758·7 the characteristic is 3
            $\log_{10}$ 375·87 ,,   ,,   ,, 2
            $\log_{10}$ 37·587 ,,   ,,   ,, 1

LOGARITHMS 45

Thus the characteristic may always be determined by inspection, and consequently is not given in the tables. This is one advantage of having 10 for a base.

34. Mantissa of a logarithm.

The decimal part of a logarithm is called the mantissa.

In general the mantissa can be calculated to any required number of figures, by the use of higher mathematics. In most tables, such as those given in this volume, the mantissa is calculated to *four* places of decimals approximately. In *Chamber's " Book of Tables "* they are calculated to seven places of decimals.

The mantissa alone is given in the tables, and the following example will show the reason why:

$$\log_{10} 168 \cdot 3 = 2 \cdot 2261$$
$$\therefore \quad 168 \cdot 3 = 10^{2 \cdot 2261}$$
$$\therefore \quad 168 \cdot 3 \div 10 = 10^{2 \cdot 2261} \div 10^1$$
$$\therefore \quad 16 \cdot 83 = 10^{2 \cdot 2261 - 1} \quad \text{(second law of indices)}$$
$$= 10^{1 \cdot 2261}$$
$$\therefore \quad \log_{10} 16 \cdot 83 = 1 \cdot 2261$$
$$\text{Similarly} \quad \log_{10} 1 \cdot 683 = 0 \cdot 2261$$
$$\text{and} \quad \log_{10} 1683 = 3 \cdot 2261$$

Thus, if a number is multiplied or divided by a power of 10, the characteristic of the logarithm of the result is changed, but **the mantissa remains unaltered.** This may be expressed as follows:

Numbers having the same set of significant figures have the same mantissa in their logarithms.

35. To read a table of logarithms.

With the use of the above rules relating to the characteristic and mantissa of logarithms, the student should have no difficulty in reading a table of logarithms.

Below is a portion of such a table, giving the logarithms of numbers between 31 and 35.

No.	Log.	1	2	3	4	5	6	7	8	9	1	2	3	4	5	6	7	8	9
31	4914	4928	4942	4955	4969	4983	4997	5011	5024	5038	1	3	4	6	7	8	10	11	12
32	5051	5065	5079	5092	5105	5119	5132	5145	5159	5172	1	3	4	5	7	8	9	11	12
33	5185	5198	5211	5224	5237	5250	5263	5276	5289	5302	1	3	4	5	6	8	9	10	12
34	5315	5328	5340	5353	5366	5378	5391	5403	5416	5428	1	3	4	5	6	8	9	10	11
35	5441	5453	5465	5478	5490	5502	5514	5527	5539	5551	1	2	4	5	6	7	9	10	11

(1)   (2)

The figures in column 1 in the complete table are the numbers from 1 to 99. The corresponding number in column 2 is the mantissa of the logarithm. As previously stated, the characteristic is not given, but can be written down by inspection. Thus $\log_{10} 31 = 1 \cdot 4914$, $\log_{10} 310 = 2 \cdot 4914$, etc. If the number has a *third significant figure*, the mantissa will be found in the appropriate column of the next nine columns.

Thus
$$\log_{10} 31 \cdot 1 = 1 \cdot 4928,$$
$$\log_{10} 31 \cdot 2 = 1 \cdot 4942, \text{ and so on}$$

If the number has a *fourth significant figure* space does not allow us to give the whole of the mantissa. But the next nine columns of what are called " mean differences " give us for every fourth significant figure a number which must be added to the mantissa already found for the first three significant figures. Thus if we want $\log_{10} 31 \cdot 67$, the mantissa for the first three significant figures 316 is 0·4997. For the fourth significant figure 7 we find in the appropriate column of mean differences the number 10. This is added to 0·4997 and so we obtain for the mantissa 5007.

$$\therefore \quad \log_{10} 31 \cdot 67 = 1 \cdot 5007$$

Anti-logarithms.

The student is usually provided with a table of anti-logarithms which contains the *numbers corresponding to given logarithms*. These could be found from a table of logarithms but it is quicker and easier to use the anti-logarithms.

The tables are similar in their use to those for logarithms, but we must remember:

(1) That the mantissa of the log only is used in the table.

(2) When the significant figures of the number have been obtained, the student must proceed to fix the decimal point in them by using the rules which we have considered for the characteristic.

Example. *Find the number whose logarithm is 2·3714.*

First using the mantissa—viz., 0·3714—we find from the anti-logarithm table that the number corresponding is given as 2352. These are the first four significant figures of the number required.

Since the characteristic is 2, the number must lie between 100 and 1000 (see § 33) and therefore it must have 3 significant figures in the integral part.

$\therefore$ The number is 235·2.

Note.—As the log tables which will be usually employed by the beginner are all calculated to base 10, the base in further work will be omitted when writing down logarithms. Thus we shall write log 235·2 = 2·3714, the base 10 being understood.

Exercise 3.

1. Write down the characteristics of the logarithms of the following numbers:

> 15, 1500, 31,672, 597, 8, 800,000
> 51·63, 3874·5, 2·615, 325·4

2. Read from the tables the logarithms of the following numbers:

> (1) 5, 50, 500, 50,000.
> (2) 4·7, 470, 47,000.
> (3) 52·8, 5·28, 528.
> (4) 947·8, 9·478, 94,780.
> (5) 5·738, 96·42, 6972.

3. Find, from the tables, the numbers of which the following are the logarithms:

> (1) 2·65, 4·65, 1·65.
> (2) 1·943, 3·943, 0·943.
> (3) 0·6734, 2·6734, 5·6734.
> (4) 3·4196, 0·7184, 2·0568.

36. Rules for the use of logarithms.

In using logarithms for calculations we must be guided by the laws which govern operations with them. Since logarithms are indices, these laws must be the same in principle as those of indices. These rules are given below; formal proofs are omitted.

(1) Logarithm of a product.

The logarithm of the product of two or more numbers is equal to the sum of the logarithms of these numbers (see first law of indices).

Thus if p and q be any numbers

$$\log (p \times q) = \log p + \log q$$

(2) Logarithm of a quotient.

The logarithm of p divided by q is equal to the logarithm of p diminished by the logarithm of q (see second law of indices).

Thus $$\log (p \div q) = \log p - \log q$$

(3) Logarithm of a power.

The logarithm of a power of a number is equal to the logarithm of the number multiplied by the index of the power (see third law of indices).

Thus $$\log a^n = n \log a$$

(4) Logarithm of a root.

This is a special case of the above (3).

Thus
$$log \ ^n\!\sqrt{a} = \log a^{\frac{1}{n}}$$
$$= \frac{1}{n} log \ a$$

37. Examples of the use of logarithms.

Example 1. *Find the value of 57·86 × 4·385.*

		No.	log.
Let	$x = 57{\cdot}86 \times 4{\cdot}385$	57·86	1·7624
Then	$\log x = \log 57{\cdot}86 + \log 4{\cdot}385$	4·385	0·6420
	$= 1{\cdot}7624 + 0{\cdot}6420$		
	$= 2{\cdot}4044$	253·7	2·4044
	$= \log 253{\cdot}7$		
$\therefore$	$x = \mathbf{253{\cdot}7}$		

Notes.—(1) The student should remember that the logs in the tables are correct to four significant figures only. Consequently he cannot be sure of four significant figures in the answer. It would be more correct to give the above answer as 254, correct to three significant figures.

(2) The student is advised to adopt some systematic way of arranging the actual operations with logarithms. Such a method is shown above.

Example 2. *Find the value of*

$$\frac{5{\cdot}672 \times 18{\cdot}94}{1{\cdot}758}.$$

		No.	log.
Let	$x = \dfrac{5{\cdot}672 \times 18{\cdot}94}{1{\cdot}758}$	5·672	0·7538
$\therefore$	$\log x = \log 5{\cdot}672 + \log 18{\cdot}94 - \log 1{\cdot}758$	18·94	1·2774
	$= 0{\cdot}7538 + 1{\cdot}2774 - 0{\cdot}2450$		2·0312
	$= 1{\cdot}7862$	1·758	0·2450
	$= \log 61{\cdot}12$		
$\therefore$	$x = 61{\cdot}12$	61·12	1·7862
or	$x = \mathbf{61{\cdot}1}$ (to three significant figures)		

Example 3. *Find the fifth root of 721·8.*

Let
$$x = \sqrt[5]{721\cdot8}$$
$$= (721\cdot8)^{\frac{1}{5}}$$

Then
$$\log x = \tfrac{1}{5} \log 721\cdot8 \text{ (see § 36(4))}$$
$$= \tfrac{1}{5}(2\cdot8584)$$
$$= 0\cdot5717$$
$$\therefore \quad x = 3\cdot730$$

Exercise 4.

Use logarithms to find the values of the following:

1. $23\cdot4 \times 14\cdot73$.
2. $43\cdot97 \times 6\cdot284$.
3. $987\cdot4 \times 1\cdot415$.
4. $42\cdot7 \times 9\cdot746 \times 14\cdot36$.
5. $28\cdot63 \div 11\cdot95$.
6. $43\cdot97 \div 6\cdot284$.
7. $23\cdot4 \div 14\cdot73$.
8. $927\cdot8 \div 4\cdot165$.
9. $94\cdot76 \times 4\cdot195 \div 27\cdot94$.
10. $\dfrac{15\cdot36 \times 9\cdot47 \times 11\cdot48}{5\cdot632 \times 21\cdot85}$.
11. $(9\cdot478)^3$.
12. $(51\cdot47)^2$.
13. $(1\cdot257)^5$.
14. $(15\cdot23)^2 \times 3\cdot142$.
15. $(5\cdot98)^2 \div 16\cdot47$.
16. $\dfrac{(91\cdot5)^2}{4\cdot73 \times 16\cdot92}$.
17. $\dfrac{(8\cdot97)^2 \times (1\cdot059)^3}{57\cdot7}$.
18. $\dfrac{4798}{(56\cdot2)^2 \div (9\cdot814)^3}$.
19. $\sqrt[4]{3\cdot417}$.
20. $\sqrt[3]{4\cdot872}$.
21. $\sqrt[3]{1\cdot625^2 \times 4\cdot738}$.
22. $\sqrt[5]{61\cdot5 \times 2\cdot73}$.
23. If $\pi r^2 = 78\cdot6$ find r when $\pi = 3\cdot142$.
24. If $\tfrac{4}{3}\pi r^3 = 15\cdot5$, find r when $\pi = 3\cdot142$.

38. Logarithms of numbers between 0 and 1.

In § 33 we gave examples of powers of 10 when the index is a positive integer. We will now consider cases in which the indices are negative.

Thus
$$10^1 = 10 \qquad \therefore \log_{10} 10 = 1$$
$$10^0 = 1 \qquad \therefore \log_{10} 1 = 0$$
$$10^{-1} = \tfrac{1}{10} = 0\cdot1 \qquad \therefore \log_{10} 0\cdot1 = -1$$
$$10^{-2} = \frac{1}{10^2} = 0\cdot01 \qquad \therefore \log_{10} 0\cdot01 = -2$$
$$10^{-3} = \frac{1}{10^3} = 0\cdot001 \qquad \therefore \log_{10} 0\cdot001 = -3$$

etc.

From these results we may deduce that:

The logarithms of numbers between 0 and 1 are always negative.

We have seen (§ 34) that if a number be divided by 10, we obtain the log of the result by subtracting 1.

Thus if   log 49·8   = 1·6972
               log 4·98   = 0·6972
               log 0·498   = 0·6972 − 1
               log 0·0498   = 0·6972 − 2
               log 0·00498 = 0·6972 − 3
From the above log 0·498 = 0·6972 − 1
                         = − 0·3028

Now, in the logs of numbers greater than unity, the mantissa remains the same when the numbers are multiplied or divided by powers of 10 (see § 34), *i.e.* with the same significant figures we have the same mantissa.

It would clearly be a great advantage if we could find a system which would enable us to use this rule for numbers less than unity, and so avoid, for example, having to write

$$\log 0·498 \text{ as } − 0·3028$$

This can be done by not carrying out the subtraction as shown above, and writing down the characteristic as negative. But to write log 0·498 as 0·6972 − 1 would be awkward. Accordingly we adopt the notation $\bar{1}$·6972 writing the minus sign above the characteristic.

It is very important to remember that

$$\bar{1}·6972 = − 1 + 0·6972$$

Thus in logarithms written in this way **the characteristic is negative and the mantissa is positive.**

With this notation   log 0·0498   = $\bar{2}$·6972
                     log 0·00498   = $\bar{3}$·6972
                     log 0·000498 = $\bar{4}$·6972 etc.

Note. —The student should note that *the negative characteristic is numerically one more than the number of zeros after the decimal point.*

Example I. *From the tables find the logs of 0·3185, 0·03185 and 0·003185.*

Using the portion of the tables in § 35, we see that the mantissa for 0·3185 will be 0·5031.

Also the characteristic is − 1.

          ∴   log 0·3185   = $\bar{1}$·5031
Similarly   log 0·03185   = $\bar{2}$·5031
and   log 0·003185 = $\bar{3}$·5031

Example 2. *Find the number whose log. is $\bar{3}$·5416.*

From the anti-log tables we find that the significant figures of the number whose mantissa is 5416 are 3480. As the characteristic is -3, there will be two zeros after the decimal point.

$\therefore$ the number is **0·003480**

Exercise 5.

1. Write down the logarithms of:

    (1) 2·798, 0·2798, 0·02798.
    (2) 4·264, 0·4264, 0·004264.
    (3) 0·009783, 0·0009783, 0·9783.
    (4) 0·06451, 0·6451, 0·0006451.

2. Write down the logarithms of:

    (1) 0·05986.   (4) 0·00009275.
    (2) 0·000473.   (5) 0·5673.
    (3) 0·007963.   (6) 0·07986.

3. Find the numbers whose logarithms are:

    (1) $\bar{1}$·3342.   (4) $\bar{4}$·6437.
    (2) $\bar{3}$·8724.   (5) $\bar{1}$·7738.
    (3) $\bar{2}$·4871.   (6) $\bar{8}$·3948.

39. Operations with logarithms which are negative.

Care is needed in dealing with the logarithms of numbers which lie between 0 and 1, since they are negative and, as shown above, are written with the characteristic negative and the mantissa positive.

A few examples will show the methods of working.

Example 1. *Find the sum of the logarithms*:

$$\bar{1}·6173, \quad \bar{2}·3415, \quad \bar{1}·6493, \quad 0·7374$$

Arranging thus
$$\begin{array}{r} \bar{1}·6173 \\ \bar{2}·3415 \\ \bar{1}·6493 \\ 0·7374 \\ \hline \bar{2}·3455 \end{array}$$

The point to be specially remembered is that the 2 which is carried forward from the addition of the mantissæ is positive, since they are positive. Consequently the addition of the characteristics becomes

$$-1 - 2 - 1 + 0 + 2 = -2$$

Example 2. *From the logarithm* $\bar{1}·6175$ *subtract the* log $\bar{3}·8463$.

$$\begin{array}{r} \bar{1}·6175 \\ \bar{3}·8463 \\ \hline 1·7712 \end{array}$$

Here in " borrowing " to subtract the 8 from the 6, the
− 1 in the top line becomes − 2, consequently on subtracting the characteristics we have

$$- 2 - (- 3) = - 2 + 3 = + 1$$

Example 3. *Multiply $\bar{2}\cdot 8763$ by 3.*

$$\begin{array}{r} \bar{2}\cdot 8763 \\ 3 \\ \hline \bar{4}\cdot 6289 \end{array}$$

From the multiplication of the mantissa, 2 is carried
forward. But this is positive and as $(- 2) \times 3 = - 6$,
the characteristic becomes $- 6 + 2 = - 4$.

Example 4. *Multiply $\bar{1}\cdot 8738$ by $1\cdot 3$.*

In a case of this kind it is better to multiply the characteristic and mantissa separately and add the results.

Thus
$$0\cdot 8738 \times 1\cdot 3 = 1\cdot 13594$$
$$- 1 \times 1\cdot 3 = - 1\cdot 3$$

$- 1\cdot 3$ is wholly negative and so we change it to $\bar{2}\cdot 7$, to
make the mantissa positive.

Then the product is the sum of

$$\begin{array}{r} 1\cdot 13594 \\ \bar{2}\cdot 7 \\ \hline \bar{1}\cdot 83594 \end{array}$$

$$\text{or} \quad \bar{1}\cdot 8359 \text{ approx.}$$

Example 5. *Divide $\bar{5}\cdot 3716$ by 3.*

Here the difficulty is that on dividing $\bar{5}$ by 3 there is a
remainder 2 which is negative, and cannot therefore be
carried on to the positive mantissa. To get over the
difficulty we write:

$$- 5 = - 6 + 1$$
or the log as
$$- 6 + 1\cdot 3716$$

Then the division of the − 6 gives us − 2 and the division
of the positive part $1\cdot 3716$ gives $0\cdot 4572$, which is positive.
Thus the complete quotient is $\bar{2}\cdot 4572$. The work might be
arranged thus:

$$\begin{array}{r} 3)\overline{\bar{6} + 1\cdot 3716} \\ \bar{2} + 0\cdot 4572 \\ \hline = \bar{2}\cdot 4572 \end{array}$$

Exercise 6.

1. Add together the following logarithms:

    (1) $\bar{2} \cdot 5178 + 1 \cdot 9438 + 0 \cdot 6138 + \bar{5} \cdot 5283$.
    (2) $3 \cdot 2165 + \bar{3} \cdot 5189 + \bar{1} \cdot 3297 + \bar{2} \cdot 6475$.

2. Find the values of:

    (1) $4 \cdot 2183 - 5 \cdot 6257$.   (3) $\bar{1} \cdot 6472 - \bar{1} \cdot 9875$.
    (2) $0 \cdot 3987 - \bar{1} \cdot 5724$.   (4) $\bar{2} \cdot 1085 - \bar{5} \cdot 6271$.

3. Find the values of:

    (1) $\bar{1} \cdot 8732 \times 2$.   (4) $\bar{1} \cdot 5782 \times 1 \cdot 5$.
    (2) $\bar{2} \cdot 9456 \times 3$.   (5) $\bar{2} \cdot 9947 \times 0 \cdot 8$.
    (3) $\bar{1} \cdot 5782 \times 5$.   (6) $\bar{2} \cdot 7165 \times 2 \cdot 5$.

4. Find the values of:

    (1) $\bar{3} \cdot 9778 \times 0 \cdot 65$.   (4) $2 \cdot 1342 \times - 0 \cdot 4$.
    (2) $\bar{2} \cdot 8947 \times 0 \cdot 84$.   (5) $1 \cdot 3164 \times - 1 \cdot 5$.
    (3) $\bar{1} \cdot 6257 \times 0 \cdot 6$.   (6) $\bar{1} \cdot 2976 \times - 0 \cdot 8$.

5. Find the values of:

    (1) $\bar{1} \cdot 4798 \div 2$.   (4) $\bar{3} \cdot 1195 \div 2$.
    (2) $\bar{2} \cdot 5637 \div 5$.   (5) $\bar{1} \cdot 6173 \div 1 \cdot 4$.
    (3) $\bar{4} \cdot 3178 \div 3$.   (6) $\bar{2} \cdot 3178 \div 0 \cdot 8$.

Use logarithms to find the values of the following:

6. $15 \cdot 62 \times 0 \cdot 987$.

7. $0 \cdot 4732 \times 0 \cdot 694$.

8. $0 \cdot 513 \times 0 \cdot 0298$.

9. $75 \cdot 94 \times 0 \cdot 0916 \times 0 \cdot 8194$.

10. $9 \cdot 463 \div 15 \cdot 47$.

11. $0 \cdot 9635 \div 29 \cdot 74$.

12. $27 \cdot 91 \div 569 \cdot 4$.

13. $0 \cdot 0917 \div 0 \cdot 5732$.

14. $5 \cdot 672 \times 14 \cdot 83 \div 0 \cdot 9873$.

15. $(0 \cdot 9173)^2$.

16. $(0 \cdot 4967)^3$.

17. $\sqrt[3]{1 \cdot 715.}$

18. $647 \cdot 2 \div (3 \cdot 715)^3$.

19. $\frac{1}{2}(48 \cdot 62)^{\frac{1}{3}}$.

20. $\sqrt[3]{\dfrac{9 \cdot 728}{3 \cdot 142}}$.

21. $(1 \cdot 697)^{2 \cdot 4}$.

22. $(19 \cdot 72)^{0 \cdot 57}$.

23. $(0 \cdot 478)^{3 \cdot 1}$.

24. $(5 \cdot 684)^{-1 \cdot 12}$.

25. $(0 \cdot 5173)^{-3 \cdot 4}$.

CHAPTER III

THE TRIGONOMETRICAL RATIOS

THE TANGENT

40. ONE of the earliest examples that we know in history of the practical applications of Geometry was the problem of finding the height of one of the Egyptian pyramids. This was solved by Thales, the Greek philosopher and mathematician who lived about 640 B.C. to 550 B.C. For this purpose he used the property of similar triangles which is stated in § 15 and he did it in this way.

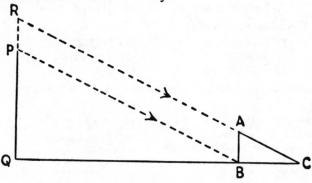

FIG. 36.

He observed the length of the shadow of the pyramid and, at the same time, that of a stick, AB, placed vertically into the ground at the end of the shadow of the pyramid (Fig. 36). QB represents the length of the shadow of the pyramid, and BC that of the stick. Then he said "*The height of the pyramid is to the length of the stick, as the length of the shadow of the pyramid is to the length of the shadow of the stick.*"

i.e. in Fig. 36,

$$\frac{PQ}{AB} = \frac{QB}{BC}.$$

Then QB, AB, and BC being known we can find PQ.

We are told that the king, Amasis, was amazed at this application of an abstract geometrical principle to the solution of such a problem.

The principle involved is practically the same as that employed in modern methods of solving the same problem. It will be well, therefore, to examine it more closely.

We note first that it is assumed that the sun's rays are parallel over the limited area involved; this assumption is justified by the great distance of the sun.

In Fig. 36 it follows that the straight lines RC and PB which represent the rays falling on the tops of the objects are parallel.

Consequently, from Theorem 2(1), § 9,

$$\angle PBQ = \angle ACB$$

These angles each represent the altitude of the sun (§ 24).

As $\angle$s PQB and ABC are right angles

$\triangle$s PQB, ABC are similar.

$$\therefore \quad \frac{PQ}{QB} = \frac{AB}{BC}$$

or as written above $\qquad \dfrac{PQ}{AB} = \dfrac{QB}{BC}.$

The solution is independent of the length of the stick AB because if this be changed the length of its shadow will be changed proportionally.

We therefore can make this important general deduction.

For the given angle ACB the ratio $\dfrac{AB}{BC}$ remains constant whatever the length of AB.

This ratio can therefore be calculated beforehand whatever the size of the angle ACB. If this be done there is no necessity to use the stick, because knowing the angle and the value of the ratio, when we have measured the length of QB, we can easily calculate PQ. Thus if the altitude were found to be 64° and the value of the ratio for this angle had been previously calculated to be 2·05, then we have

$$\frac{PQ}{QB} = 2\cdot05$$

and $\qquad\qquad PQ = QB \times 2\cdot05.$

41. Tangent of an angle.

The idea of a constant ratio for every angle is so important that we must examine it in greater detail.

Let POQ (Fig. 37) be any acute angle. From points A, B, C on one arm draw perpendiculars AD, BE, CF to the other arm. These being parallel,

$\qquad \angle$s OAD, OBE, OCF are equal (Theorem 2 (1))

and $\qquad \angle$s ODA, OEB, OFC are right-$\angle$s.

$\quad \therefore \quad \triangle$s AOD, BOE, COF are similar.

$$\therefore \quad \frac{AD}{OD} = \frac{BE}{OE} = \frac{CF}{OF} \quad \text{(Theorem 10, § 15)}$$

Similar results follow, no matter how many points are taken on OQ.

∴ for the angle POQ the ratio of the perpendicular drawn from a point on one arm of the angle to the distance intercepted on the other arm is constant.

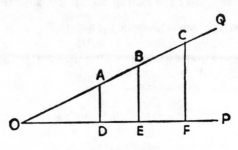

Fig. 37.

This is true for any angle; each angle has its own particular ratio and can be identified by it.

This constant ratio is called the tangent of the angle.

The name is abbreviated in use to *tan*.

Thus for $\angle POQ$ above we can write

$$\tan POQ = \frac{AD}{OD}.$$

42. Right-angled triangles.

Before proceeding further we will consider formally by means of the tangent, the relations which exist between the sides and angles of a right-angled triangle.

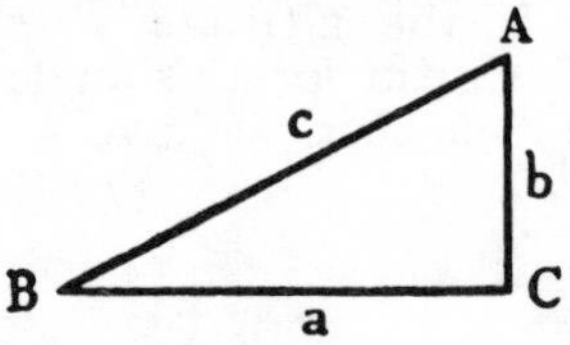

Let ABC (Fig. 38) be a right-angled triangle.

Let the sides opposite the angles be denoted by

Fig. 38.

a (opp. A), b (opp. B), c (opp. C).

(This is a general method of denoting sides of a right-angled $\triangle$.)

Then, as shown in § 41:

$$\tan B = \frac{AC}{BC} = \frac{b}{a}$$

$$\therefore \quad a \tan B = b$$

and

$$a = \frac{b}{\tan B}.$$

Thus any one of the three quantities a, b, $\tan B$ can be determined when the other two are known.

43. Notation for angles.

(1) As indicated above we sometimes, for brevity, refer to an angle by using only the *middle* letter of the three which define the angle.

Thus we use $\tan B$ for $\tan ABC$.

This must not be used when there is any ambiguity, as, for example, when there is more than one angle with its vertex at the same point.

(2) When we refer to angles in general we frequently use a Greek letter, usually θ (pronounced " theta ") or φ (pronounced " phi ") or α, β, or γ (alpha, beta, gamma).

44. Changes in the tangent in the first quadrant.

In Fig. 39 let OA a straight line of *unit* length rotate from a fixed position on OX until it reaches OY, a straight line perpendicular to OX.

From O draw radiating lines to mark $10°$, $20°$, $30°$, etc.

From A draw a straight line AM perpendicular to OX and let the radiating lines be produced to meet this.

Let OB be any one of these lines.

Then $$\tan BOA = \frac{BA}{OA}.$$

Since OA is of unit length, then the length of BA, on the scale selected, will give the actual value of $\tan BOA$.

Similarly the tangents of other angles $10°$, $20°$, etc., can be read off by measuring the corresponding intercept on AM.

If the line OC corresponding to $45°$ be drawn then $\angle ACO$ is also $45°$ and AC equals OA (Theorem 3, § 11).

$$\therefore \quad AC = 1$$
$$\therefore \quad \tan 45° = 1.$$

At the initial position, when OA is on OX the angle is $0°$, the length of the perpendicular from A is zero, and the tangent is also zero.

From examination of the values of the tangents as marked on AM, we may conclude:

(1) $\tan 0°$ is 0.
(2) As the angle increases, $\tan θ$ increases.
(3) $\tan 45° = 1$.
(4) For angles greater than $45°$, the tangent is greater than 1.

(5) As the angle approaches 90° the tangent increases very rapidly. When it is almost 90° it is clear that the radiating line will meet *AM* at a very great distance, and when it coincides with *OY* and 90° is reached, we say that the tangent has become infinitely great.

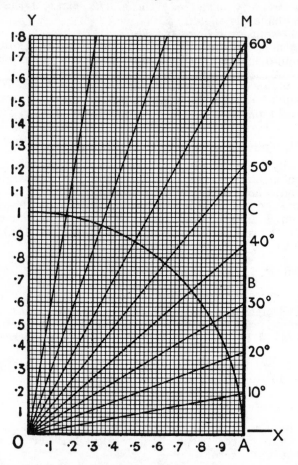

Fig. 39.

This can be expressed by saying that
As θ approaches 90°, tan θ approaches infinity.

This may be expressed formally by the notation

when $\theta \longrightarrow 90°$, $\tan \theta \longrightarrow \infty$.

The symbol ∞, commonly called infinity, means a number greater than any conceivable number.

45. A table of tangents.

Before use can be made of tangents in practical applications and calculations, it is necessary to have a table which will give with great accuracy the tangents of all angles which may be required.   It must also be possible from it to obtain the angle corresponding to a known tangent.

A rough table could be constructed by such a practical method as is indicated in the previous paragraph.   But results obtained in this way would not be very accurate.

By the methods of more advanced mathematics, however, these values can be calculated to any required degree of accuracy.   For elementary work it is customary to use tangents calculated correctly to four places of decimals. Such a table will be found at the end of this book.

A small portion of this table, giving the tangents of angles from 25° to 29° inclusive is given below, and this will serve for an explanation as to how to use it.

NATURAL TANGENTS.

Degrees.	0′	6′	12′	18′	24′	30′	36′	42′	48′	54′	Mean Differences.				
											1	2	3	4	5
25	0·4663	4684	4706	4727	4748	4770	4791	4813	4834	4856	4	7	11	14	18
26	0·4877	4899	4921	4942	4964	4986	5008	5029	5051	5073	4	7	11	15	18
27	0·5095	5117	5139	5161	5184	5206	5228	5250	5272	5295	4	7	11	15	18
28	0·5317	5340	5362	5384	5407	5430	5452	5475	5498	5520	4	8	11	15	19
29	0·5543	5566	5589	5612	5635	5658	5681	5704	5727	5750	4	8	12	15	19

(1) The first column indicates the angle in degrees.

(2) The second column states the corresponding tangent.

Thus   tan 27° = 0·5095.

(3) If the angle includes minutes we must use the remaining columns.

(a) If the number of minutes is a multiple of " 6 ", the figures in the corresponding column gives the decimal part of the tangent.   Thus tan 25° 24′ will be found under the column marked 24′.   From this we see

tan 25° 24′ = 0·4748.

(b) If the number of minutes is not an exact multiple of 6, we use the columns headed " mean differences " for angles which are 1, 2, 3, 4, or 5 minutes more than the multiple of " 6."

Thus if we want tan 26° 38′, this being 2′ more than 26° 36′, we look under the column headed 2 in the line of 26°. The difference is 7. This is added to tan 26° 36′, *i.e.* 0·5008.

Thus   tan 26° 38′ = 0·5008 + ·0007
                          = 0·5015.

An examination of the first column in the table of tangents will show you that as the angles increase and approach 90° the tangents increase very rapidly. Consequently for angles greater than 45° the whole number part is given as well as the decimal part. For angles greater than 74° the mean differences become so large and increase so rapidly that they cannot be given with any degree of accuracy. If the tangents of these angles are required, the student must consult such a book as *Chambers' Mathematical Tables*, where *seven* significant figures are given. This book should be found in the library of everybody who is studying Trigonometry and its applications.

46. Examples of the uses of tangents.

We will now consider a few examples illustrating practical applications of tangents. The first is suggested by the problem mentioned in § 24.

Example 1. *At a point 168 m horizontally distant from the foot of a church tower, the angle of elevation of the top of the tower is 38° 15′.*
Find the height above the ground of the top of the tower.

In Fig. 40 PQ represents the height of P above the ground.

We will assume that the distance from O is represented by OQ.

Then $\angle POQ$ is the angle of elevation and equals 38° 15′.

$$\therefore \quad \frac{PQ}{OQ} = \tan 38° 15′$$

$$\therefore \quad PQ = OQ \times \tan 38° 15′$$
$$= 168 \times \tan 38° 15′$$
$$= 168 \times 0·7883$$

Taking logarithms of both sides

$$\log (PQ) = \log 168 + \log (0·7883)$$
$$= 2·2253 + \bar{1}·8967$$
$$= 2·1220$$
$$= \log 132·4$$
$$\therefore \quad PQ = 132 \text{ m approx.}$$

Example 2. *A man, who is 168 cm in height, noticed that the length of his shadow in the sun was 154 cm. What was the altitude of the sun?*

In Fig. 41 let PQ represent the man and QR represent the shadow.

Then PR represents the sun's ray and $\angle PRQ$ represents the sun's altitude.

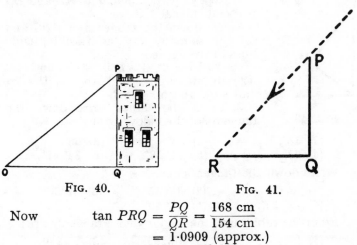

FIG. 40.   FIG. 41.

Now
$$\tan PRQ = \frac{PQ}{QR} = \frac{168 \text{ cm}}{154 \text{ cm}}$$
$$= 1{\cdot}0909 \text{ (approx.)}$$
$$= \tan 47° 29'$$

∴ the sun's altitude is 47° 29'.

Example 3. *Fig. 42 represents a section of a symmetrical roof in which AB is the span, and OP the rise. (P is the midpoint of AB.) If the span is 22 m and the rise 7 m find the slope of the roof (i.e. the angle OBA).*

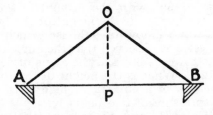

FIG. 42.

OAB is an isosceles triangle, since the roof is symmetrical.

∴ OP is perpendicular to AB (Theorem 3, § 11).

$$\therefore \quad \tan OBP = \frac{OP}{PB}$$
$$= \tfrac{7}{11} = 0{\cdot}6364 \text{ (approx.)}$$
$$= \tan 32° 28' \text{ (approx.)}$$
$$\therefore \quad \angle OBP = 32° 28'.$$

Exercise 7.

1. In Fig. 43 ABC is a right-angled triangle with C the right angle.

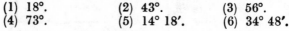

Draw CD perpendicular to AB and DQ perpendicular to CB.

Write down the tangents of ABC and CAB in as many ways as possible, using lines of the figure.

2. In Fig. 43, if AB is 15 cm and AC 12 cm in length, find the values of tan ABC and tan CAB.

Fig. 43.

3. From the tables write down the tangents of the following angles:

  (1) 18°.   (2) 43°.   (3) 56°.
  (4) 73°.   (5) 14° 18′.   (6) 34° 48′.

4. Write down the tangents of:

  (1) 9° 17′.   (2) 31° 45′.   (3) 39° 5′.
  (4) 52° 27′.   (5) 64° 40′.

5. From the tables find the angles whose tangents are:

  (1) 0·5452.   (2) 1·8265.   (3) 2·8239.
  (4) 1·3001.   (5) 0·6707.   (6) 0·2542.

6. When the altitude of the sun is 48° 24′, find the height of a flagstaff whose shadow is 7·42 m long.

7. The base of an isosceles triangle is 10 mm and each of the equal sides is 13 mm. Find the angles of the triangle.

8. A ladder rests against the top of the wall of a house and makes an angle of 69° with the ground. If the foot is 7·5 m from the wall, what is the height of the house?

9. From the top window of a house which is 1·5 km away from a tower it is observed that the angle of elevation of the top of the tower is 36° and the angle of depression of the bottom is 12°. What is the height of the tower?

10. From the top of a cliff 32 m high it is noted that the angles of depression of two boats lying in the line due east of the cliff are 21° and 17°. How far are the boats apart?

11. Two adjacent sides of a rectangle are 15·8 cms and 11·9 cms. Find the angles which a diagonal of the rectangle makes with the sides.

12. P and Q are two points directly opposite to one another on the banks of a river. A distance of 80 m is measured along one bank at right angles to PQ. From the end of this line the angle subtended by PQ is 61°. Find the width of the river.

SINES AND COSINES

47. In Fig. 44 from a point A on one arm of the angle ABC, a perpendicular is drawn to the other arm.

We have seen that the ratio $\dfrac{AC}{BC} = \tan ABC$.

Now let us consider the ratios of each of the lines AC and BC to the hypotenuse AB.

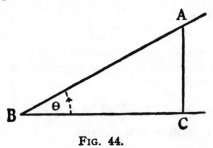

FIG. 44.

(1) The *ratio* $\dfrac{AC}{AB}$, *i.e.* the ratio of the side opposite to the angle to the hypotenuse.

This ratio is also constant, as was the tangent, for the angle ABC, *i.e.* wherever the point A is taken, the ratio of AC to AB remains constant.

This ratio is called the sine of the angle and is denoted by $\sin ABC$.

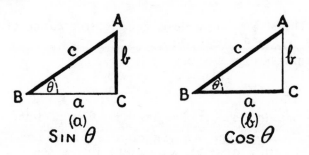

FIG. 45.

(2) The *ratio* $\dfrac{BC}{AB}$, *i.e.* the ratio of the intercept to the hypotenuse.

This ratio is also constant for the angle and is called the cosine. It is denoted by $\cos ABC$.

The beginner is sometimes apt to confuse these two

ratios. The way in which they are depicted by the use of thick lines in Fig. 45 may assist the memory. If the sides of the $\triangle ABC$ are denoted by a, b, c in the usual way and the angle ABC by θ (pronounced *theta*).

Then in 45(a) $\qquad sin\ \theta = \dfrac{b}{c}$ (1)

$\qquad$ 45(b) $\qquad cos\ \theta = \dfrac{a}{c}$ (2)

From (1) we get $\qquad b = c\ sin\ \theta$
,, (2) ,, ,, $\qquad a = c\ cos\ \theta$

Since in the fractions representing $sin\ \theta$ and $cos\ \theta$ above, the denominator is the hypotenuse, which is the greatest side of the triangle, then

$sin\ \theta$ and $cos\ \theta$ cannot be greater than unity.

48. Ratios of complementary angles.

In Fig. 45, since $\angle C$ is a right angle.

$$\therefore \quad \angle A + \angle B = 90°$$

$\therefore$ $\angle A$ and $\angle B$ are complementary (see § 7).

Also $\qquad \sin A = \dfrac{a}{c}$

and $\qquad \cos B = \dfrac{a}{c}$

$$\therefore \quad \sin A = \cos B$$

$\therefore$ **The sine of an angle is equal to the cosine of its complement, and vice versa.**

This may be expressed in the form:

$$\sin \theta = \cos (90° - \theta)$$
$$\cos \theta = \sin (90° - \theta)$$

49. Changes in the sines of angles in the first quadrant.

Let a line, OA, a *unit in length*, rotate from a fixed position (Fig. 46) until it describes a quadrant, that is the $\angle DOA$ is a right angle.

From O draw a series of radii to the circumference corresponding to the angles $10°, 20°, 30°, \ldots$

From the points where they meet the circumference draw lines perpendicular to OA.

Considering any one of these, say BC, corresponding to $40°$.

Then $\sin BOC = \dfrac{BC}{OB}$

But OB is of unit length.

∴ BC represents the value of sin BOC, in the scale in which OA represents unity.

Consequently the various perpendiculars which have been drawn represent the sines of the corresponding angles.

Examining these perpendiculars we see that *as the angles increase from 0° to 90° the sines continually increase.*

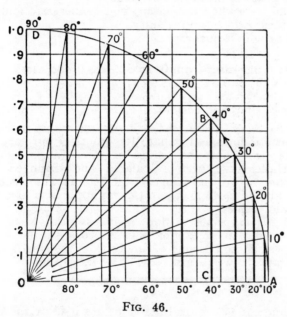

FIG. 46.

At 90° the perpendicular coincides with the radius

∴   **sin 90° = 1.**

At 0° the perpendicular vanishes.

∴   **sin 0° = 0.**

Summarising these results:

In the first quadrant

  (1) **sin 0° = 0.**
  (2) **As θ increases from 0° to 90°, sin θ increases.**
  (3) **sin 90° = 1.**

50. Changes in the cosines of angles in the first quadrant.

Referring again to Fig. 46 and considering the cosines of the angles formed as OA rotates, we have as an example

$$\cos BOC = \frac{OC}{OB}.$$

As before, OB is of unit length.

∴ OC represents in the scale taken, cos BOC.

Consequently the lengths of these intercepts on OA represent the cosines of the corresponding angles.

These *decrease as the angle increases*.

When 90° is reached this intercept becomes zero and at 0° it coincides with OA and is unity.

Hence in the first quadrant

(1) **cos 0° = I.**
(2) **As θ increases from 0° to 90°, cos θ decreases.**
(3) **cos 90° = 0.**

51. Tables of sines and cosines.

As in the case of the tangent ratio, it is necessary in order to make use of sines and cosines for practical purposes to compile tables giving the values of these ratios for all angles. These have been calculated and arranged by methods similar to the tangent tables and the general directions given in § 45 for their use will apply also to those for sines and cosines.

The table for cosines is not really essential when we have the tables of sines, for since cos θ = sin (90° − θ) (see § 48) we can find cosines of angles from the sine table.

For example, if we require cos 47°, we know that

$$\cos 47° = \sin (90° - 47°)$$
$$= \sin 43°.$$

∴   to find cos 47° we read the value of sin 43° in the sine table.

In practice this process takes longer and is more likely to lead to inaccuracies than finding the cosine direct from a table. Consequently separate tables for cosines are included at the end of this book.

There is one difference between the sine and cosine tables which the student must remember when using them.

As we have seen in § 50, as angles in the first quadrant increase, *sines increase but cosines decrease*. Therefore when using the columns of mean differences for *cosines these differences must be subtracted*.

52. Examples of the use of sines and cosines.

Example I. *The length of each of the legs of a pair of ladders is 2·5 m. The legs are opened out so that the distance between the feet is 2 m. What is the angle between the legs?*

In Fig. 47, let AB, AC represent the legs of the ladders. These being equal, BAC is an isosceles triangle.

∴ AO the perpendicular to the base BC, from the vertex bisects the vertical angle BAC, and also the base.

$$\therefore \quad BO = OC = 1 \text{ m}$$

We require to find the angle BAC.

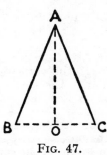

FIG. 47.

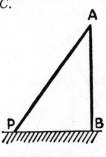

FIG. 48.

Now $\sin BAO = \dfrac{BO}{BA}$

$= \dfrac{1}{2·5} = 0·4$

$= \sin 23° 35'$ (from the tables)

∴ $\angle BAO = 23° 35'$

But $\angle BAC = 2 \times \angle BAO$

∴ $\angle BAC = 2 \times 23° 35'$

$= 47° 10'$.

Example 2. *A 30 m ladder on a fire engine has to reach a window 26 m from the ground which is horizontal and level. What angle, to the nearest degree, must it make with the ground and how far from the building must it be placed?*

Let AB (Fig. 48) represent the height of the window at A above the ground.

Let AP represent the ladder.

To find $\angle APB$ we may use its sine for

$\text{Sin } APB = \dfrac{AB}{AP} = \dfrac{26}{30}$

$= 0·8667$

$= \sin 60° 5'$ (from the tables)

∴ $APB = 60° 5'$

$= 60°$ **(to nearest degree)**.

To find PB we use the cosine of APB

for $$\cos APB = \frac{PB}{AP}$$

$$\therefore \quad PB = AP \cos APB$$
$$= 30 \times \cos 60^\circ\ 5'$$
$$= 30 \times 0\cdot5013$$
$$= 15\cdot04$$
$$\therefore \quad \boldsymbol{PB} = 15 \text{ m approx.}$$

Example 3. *The height of a cone is 18 cm and the angle at the vertex is 88°. Find the slant height.*

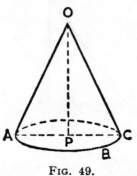

FIG. 49.

Let $OABC$ (Fig. 49) represent the cone, the vertex being O and ABC the base.

Let the $\triangle OAC$ represent a section through the vertex O and perpendicular to the base.

It will be an isosceles triangle and P the centre of its base will be the foot of the perpendicular from O to the base.

OP will also bisect the vertical angle AOC (Theorem 3).

OP represents the height of the cone and is equal to 18 cm.

OC represents the slant height.

Now $$\cos POC = \frac{OP}{OC}$$

$$\therefore \quad OP = OC \cos POC$$
$$\therefore \quad OC = OP \div \cos POC$$
$$= 18 \div \cos 44^\circ$$
$$= 18 \div 0\cdot7193$$

Taking logs: $$\log (OC) = \log 18 - \log 0\cdot7193$$
$$= 1\cdot2553 - \bar{1}\cdot8569$$
$$= 1\cdot3984$$
$$= \log 25\cdot02$$
$$\therefore \quad \boldsymbol{OC} = 25 \text{ cm approx.}$$

Example 4. *Fig. 50 represents a section of a symmetrical roof frame. $PA = 28\,m$, $AB = 6\,m$, $\angle OPA = 21^\circ$; find OP and OA.*

(1) We can get OP if we find $\angle OPB$. To do this we must first find $\angle APB$.

$$\sin APB = \frac{AB}{AP} = \frac{6}{28} = 0\cdot2143 = \sin 12^\circ\ 23'.$$
$$\therefore \quad \angle OPB = \angle OPA + \angle APB$$
$$= 21^\circ + 12^\circ\ 23' = 33^\circ\ 23'$$

Next find PB, which divided by OP gives cos OPB.

$$PB = AP \cos APB = 28 \cos 12° \ 23'$$
$$= 28 \times 0.9768$$
$$= 27.35 \text{ approx.}$$

Note.—We could also use the Theorem of Pythagoras.

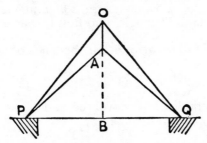

FIG. 50.

Now $\dfrac{PB}{OP} = \cos OPB$

$\therefore$ $OP = PB \div \cos OPB$
$\therefore$ $OP = 27.35 \div \cos 33° \ 23'$
$\qquad\qquad = 27.35 \div 0.8350$
$\therefore \ \log OP = \log 27.35 - \log 0.8350$
$\qquad\qquad = 1.4370 - \bar{1}.9217 = 1.5153$
$\qquad\qquad = \log 32.75$
$\therefore \ \ \textbf{OP} = 32.75 \text{ m.}$

(2) *To find OA.* This is equal to $OB - AB$.
We must therefore find OB.

Now $\dfrac{OB}{OP} = \sin OPB$

$\therefore$ $OB = OP \sin OPB$
$\qquad\qquad = 32.75 \times \sin 33° \ 23'$
$\qquad\qquad = 32.75 \times 0.5503$
$\therefore \ \log OB = \log 32.75 + \bar{1}.7406 = 1.2559$
$\qquad\qquad = \log 18.03$
$\therefore \ \ OB = 18.03$

and $OA = OB - AB$
$\qquad\qquad = 18.03 - 6$
$\qquad\qquad = 12.03 \text{ m.}$

Exercise 8.

1. Using the triangle of Fig. 43 write down in as many
ways as possible (1) the sines, (2) the cosines, of $\angle ABC$ and
$\angle CAB$, using the lines of the figure.

2. Draw a circle with radius 45 mm. Draw a chord of length 60 mm. Find the sine and cosine of the angle subtended by this chord at the centre.

3. In a circle of 4 cm radius a chord is drawn subtending an angle of 80° at the centre. Find the length of the chord and its distance from the centre.

4. The sides of a triangle are 135 mm, 180 mm, and 225 mm. Draw the triangle, and find the sines and cosines of the angles.

5. From the tables write down the sines of the following angles:

      (1) 14° 36'.   (2) 47° 44'.   (3) 69° 17'.

6. From the tables write down the angles whose sines are:

      (1) 0·4970.   (2) 0·5115.   (3) 0·7906.

7. From the tables write down the cosines of the following angles:

      (1) 20° 46'.   (2) 44° 22'.   (3) 62° 39'.
      (4) 38° 50'.   (5) 79° 16'.   (6) 57° 23'.

8. From the tables write down the angles whose cosines are:

      (1) 0·5332.   (2) 0·9358.   (3) 0·3546.
      (4) 0·2172.   (5) 0·7910.   (6) 0·5140.

9. A certain uniform incline rises 10·5 km in a length of 60 km along the incline. Find the angle between the incline and the horizontal.

10. In a right-angled triangle the sides containing the right angle are 4·5 m and 5·8 m. Find the angles and the length of the hypotenuse.

11. In the diagram of a roof frame shown in Fig. 42, find the angle at which the roof is sloped to the horizontal when $OP = 1·3$ m and $OB = 5·4$ m.

12. A rope 65 m long is stretched out from the top of a flagstaff 48 m high to a point on the ground which is level. What angle does it make with the ground and how far is this point from the foot of the flagstaff?

53. Cosecant, secant and cotangent.

From the reciprocals of the sine, cosine and tangent we can obtain three other ratios connected with an angle, and problems frequently arise where it is more convenient to employ these instead of using the reciprocals of the original ratios.

These reciprocals are called the *cosecant, secant,* and *cotangent* respectively, abbreviated to cosec, sec and cot.

Thus
$$\text{cosec } \theta = \frac{1}{\sin \theta}$$
$$\text{sec } \theta = \frac{1}{\cos \theta}$$
$$\text{cot } \theta = \frac{1}{\tan \theta}$$

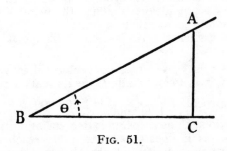

FIG. 51.

These can be expressed in terms of the sides of a right-angled triangle with the usual construction (Fig. 51) as follows:

$$\frac{AC}{AB} = \sin \theta, \quad \frac{AB}{AC} = \text{cosec } \theta$$
$$\frac{BC}{AB} = \cos \theta, \quad \frac{AB}{BC} = \text{sec } \theta$$
$$\frac{AC}{BC} = \tan \theta, \quad \frac{BC}{AC} = \text{cot } \theta$$

Ratios of complementary angles.

In continuation of § 48 we note that:

since
$$\tan ABC = \frac{AC}{BC}$$
and
$$\cot BAC = \frac{AC}{BC}$$
$$\therefore \quad \tan \theta = \cot (90° - \theta)$$

or the *tangent of an angle is equal to the cotangent of its complement.*

54. Changes in the reciprocal ratios of angles in the first quadrant.

The changes in the values of these ratios can best be examined by reference to the corresponding changes in the values of their reciprocals (see §§ 44, 49 and 50 in this chapter).

The following general relations between a ratio and its reciprocal should be noted:

(a) *When the ratio is increasing its reciprocal is decreasing,* and vice versa.

(b) *When a ratio is a maximum its reciprocal will be a minimum,* and vice versa.

Consequently since the *maximum* value of the sine and cosine in the first quadrant is unity, the *minimum* value of the cosecant and secant must be unity.

(c) The case when a ratio is zero needs special examination.

If a number is very large, its reciprocal is very small. Conversely if it is very small its reciprocal is very large. Thus the reciprocal of $\frac{1}{1,000,000}$ is 1,000,000.

When a ratio such as a cosine is decreasing until it finally becomes zero, as it does when the angle reaches 90°, the secant approaches infinity. With the notation employed in § 44 this can be expressed as follows.

As $\theta \longrightarrow 90°$, sec $\theta \longrightarrow \infty$.

55. Changes in the cosecant.

Bearing in mind the above, and remembering the changes in the sine in the first quadrant as given in § 49.
(1) Cosec 0° is infinitely large.
(2) As θ *increases* from 0° to 90°, cosec θ *decreases*.
(3) cosec 90° = 1.

56. Changes in the secant.

Comparing with the corresponding changes in the cosine we see:
(1) sec 0° = 1.
(2) As θ *increases* from 0 to 90°, sec θ *increases*.
(3) As $\theta \longrightarrow 90°$, sec $\theta \longrightarrow \infty$.

57. Changes in the cotangent.

Comparing the corresponding changes of the tan θ as given in § 44 we conclude:
(1) As $\theta \longrightarrow 0°$, cot $\theta \longrightarrow \infty$.
(2) As θ *increases*, cot θ *decreases*.
(3) cot 45° = 1.
(4) As $\theta \longrightarrow 90°$, cot $\theta \longrightarrow 0$.

58. Graphs of the trigonometrical ratios.

In Figs. 52, 53, 54 are shown the graphs of sin θ, cos θ and tan θ respectively for angles in the first quadrant. The student should draw them himself, if possible, on squared paper, obtaining the values either by the graphical methods suggested in Figs. 39 and 46 or from the tables.

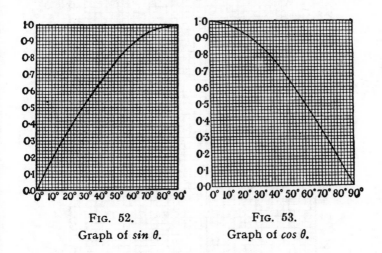

Fig. 52.

Graph of *sin θ*.

Fig. 53.

Graph of *cos θ*.

59. Logarithms of trigonometrical ratios.

Calculations in trigonometry are shortened and obtained more accurately by the use of tables giving the logarithms of sines, cosines and tangents. The advantage of their use can be illustrated by the following examples.

Find the value of sin 57° × tan 24°.

(1) We might proceed as follows.

Let $$x = \sin 57° \times \tan 24°$$
$$= 0.8387 \times 0.4452.$$

Taking logs, $\log x = \log (0.8387) + \log (0.4452)$
$$= \bar{1}.9236 + \bar{1}.6486$$

and then we proceed as usual.

This method involves the use of two sets of tables.

    (*a*) Tables of trigonometrical ratios.
    (*b*) Logarithms.

Instead of thus using two sets of tables we can use the tables which give directly the logarithms of the trigonometrical ratios.

These are the tables at the end of the book, headed,

Logarithms of sines
Logarithms of cosines
Logarithms of tangents.

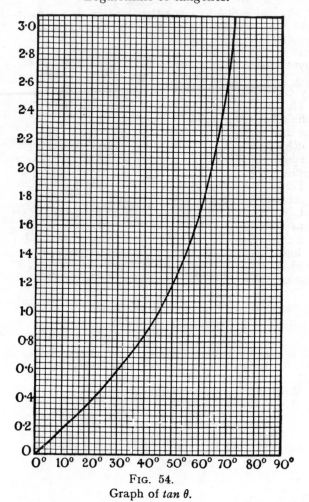

FIG. 54.

Graph of *tan θ*.

(2) The following is the solution of the above problem, using these tables.

Let
$$x = \sin 57° \times \tan 24°$$
$$\therefore \log x = \log \sin 57° + \log \tan 24°$$
$$= \bar{1} \cdot 9236 + \bar{1} \cdot 6486.$$

and so we reach the same conclusion as above in *one* step instead of two.

60. Characteristics of " logarithms of sines ", etc.

The student may find some difficulty at first in using these tables of logarithmic sines, etc., on account of the characteristics. As we have seen, all sines and cosines and tangents of angles less than 45° are less than unity. Consequently the characteristics of their logarithms are always negative (see § 38).

They can be dealt with in two ways:

(1) The characteristic may be printed in the first column, as in the tables in this book.

Thus log (sin 20°) is written $\bar{1}\cdot5341$.

In other columns the mantissa only is printed, as with ordinary tables of logs, and the negative characteristic must be supplied by the student.

(2) To avoid printing these negative characteristics it has been a custom in most tables to add 10 to the characteristic so that log (sin 20°) would be printed as $9\cdot5341$. If such tables are used by the student his easiest plan is to subtract 10 from the characteristic when writing down the logarithm.

The logarithms of cosecants, secants and tangents are not included in the tables given in this book. The student may use instead the logarithms of their reciprocals, the sines, cosines and tangents.

For example, since

$$\sec \theta = \frac{1}{\cos \theta}$$
$$\therefore \quad \log \sec \theta = \log 1 - \log \cos \theta$$
$$= 0 - \log \cos \theta \qquad \text{(see § 33)}$$
$$= - \log \cos \theta$$

It should be noted that the logarithm of a number is equal to − (log of its reciprocal).

Note.—Before proceeding to work examples on these tables the student is advised to revise § 39 in the chapter on logarithms.

Worked Examples.

Example 1. *From a certain point the angle of elevation of the top of a church spire is found to be 11°. The guide book tells me that the height of the spire is 260 m. If I am on the same horizontal level as the bottom of the tower, how far am I away from it?*

In Fig. 55 let AB represent the tower and spire,

$$AB = 260 \text{ m.}$$

Let O be the point of observation.
We require to find OB.

Let $\qquad OB = x$

Then $\qquad \dfrac{x}{260} = \cot 11°$

$\qquad \therefore \quad x = 260 \cot 11°$ **(1)**

$\qquad \therefore \quad x = 260 \times 5\cdot1446$

$\qquad \therefore \quad \log x = \log 260 + \log 5\cdot1446$

$\qquad\qquad\qquad = 2\cdot4150 + 0\cdot7113$

$\qquad\qquad\qquad = 3\cdot1263$

$\qquad\qquad\qquad = \log 1338$

$\qquad \therefore \quad \boldsymbol{x} = 1338$ m approx.

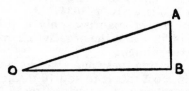

FIG. 55.

If logarithmic cotangents are used, then from (1) we get

$\qquad \log x = \log 260 + \log \cot 11°$ (or $- \log \tan 11°$)

$\qquad\qquad\quad = 2\cdot4150 + 0\cdot7113$

$\qquad\qquad\quad = 3\cdot1263$

$\qquad\qquad\quad = \log 1338$

$\qquad \therefore \quad \boldsymbol{x} = 1338$ m.

Example 2. *Find the value of $2 \sin \theta \cos \theta$ when $\theta = 38° 42'$.*

Let $\qquad x = 2 \sin \theta \cos \theta$

Then $\qquad \log x = \log 2 + \log \sin \theta + \log \cos \theta$

$\qquad\qquad\qquad = 0\cdot3010 + \bar{1}\cdot7960 + \bar{1}\cdot8923$

$\qquad\qquad\qquad = \bar{1}\cdot9893$

$\qquad \therefore \quad \boldsymbol{x} = 0\cdot9757.$

Example 3. *Find the value of $\dfrac{b - c}{b + c} \cot \dfrac{A}{2}$, when $b = 25\cdot6$, $c = 11\cdot2$, $A = 57°$.*

Since $\qquad\qquad\qquad\qquad b = 25\cdot6$

and $\qquad\qquad\qquad\qquad c = 11\cdot2$

$\qquad \therefore \quad b + c = 36\cdot8$

$\qquad\qquad\qquad\quad b - c = 14\cdot4$

and $\qquad\qquad\qquad\qquad \dfrac{A}{2} = 57° \div 2 = 28° 30'.$

Let $\qquad x = \dfrac{b - c}{b + c} \cot \dfrac{A}{2}$

Then $\qquad x = \dfrac{14 \cdot 4}{36 \cdot 8} \cot 28° 30'.$

Taking logs, $\log x = \log 14 \cdot 4 + \log \cot 28° 30' - \log 36 \cdot 8$

$\qquad\qquad = \bar{1} \cdot 8578$

$\qquad\qquad = \log 0 \cdot 7206$

$\therefore \quad x = 0 \cdot 7206$

No.	Log.
14·4	1·1584
cot 28° 30′	0·2652
	1·4236
36·8	1·5658
	1·8578

Exercise 9

1. From the tables find the following:

   (1) cosec 35° 24′.   (4) sec. 53° 5′.
   (2) cosec 59° 45′.   (5) cot 39° 42′.
   (3) sec 42° 37′.   (6) cot 70° 34′.

2. From the tables find the angle:

   (1) When the cosecant is 1·1476.
   (2) When the secant is 2·3443.
   (3) When the cotangent is 0·3779.

3. The height of an isosceles triangle is 38 mm and each of the equal angles is 52°.   Find the lengths of the equal sides.

4. Construct a triangle with sides 5 cm, 12 cm and 13 cm in length.   Find the cosecant, secant and tangent of each of the acute angles.   Hence find the angles from the tables.

5. A chord of a circle is 3 m long and it subtends an angle of 63° at the centre.   Find the radius of the circle.

6. A man walks up a steep road the slope of which is 8°. What distance must he walk so as to rise 1 km?

7. Find the values of:

   (a) $\dfrac{8 \cdot 72}{9 \cdot 83} \sin 23°.$

   (b) $\cos A \sin B$ when $A = 40°$, $B = 35°$.

8. Find the values of:

   (a) $\sin^2 \theta$ when $\theta = 28°$.

   (b) $2 \sec \theta \cot \theta$ when $\theta = 42°$.

Note.—$\sin^2 \theta$ is the usual way of writing $(\sin \theta)^2$.

9. Find the values of:

    (a) $\tan A \tan B$, when $A = 53°$, $B = 29°$.

    (b) $\dfrac{a \sin B}{b}$ when $a = 50$, $b = 27$, $B = 66°$.

10. Find the values of:

    (a) $\sec^2 43°$.   (b) $2 \cos^2 28°$.

11. Find the value of: $\sqrt{\dfrac{\sin 53° 27'}{\tan 68° 40'}}$.

12. Find the value of $\cos^2 \theta - \sin^2 \theta$.

    (1) When $\theta = 37° 25'$.   (2) When $\theta = 59°$.

13. If $\tan \dfrac{\theta}{2} = \sqrt{\dfrac{239 \times 25}{397 \times 133}}$ find θ.

14. Find the value of $2 \sin \dfrac{A + B}{2} \cos \dfrac{A - B}{2}$ when $A = 57° 14'$ and $B = 22° 29'$.

15. If $\mu = \dfrac{\sin \theta}{\cot a}$ find μ when $\theta = 10° 25'$ and $a = 28° 7'$.

16. If $A = \frac{1}{2} ab \sin \theta$, find A when $a = 28 \cdot 5$, $b = 46 \cdot 7$ and $\theta = 56° 17'$.

Some applications of trigonometrical ratios.

61. Solution of right-angled triangles.

By solving a right-angled triangle we mean, if certain sides or angles are given we require to find the remaining sides and angles.

Right-angled triangles can be solved:

(1) By using the appropriate trigonometrical ratios.

(2) By using the Theorem of Pythagoras (see Theorem **9**, § 14).

We give a few examples.

(a) **Given the two sides which contain the right angle.**

To solve this:

(1) The other angles can be found by the tangent ratios.

(2) The hypotenuse can be found by using secants and cosecants, or the Theorem of Pythagoras.

Example I. *Solve the right-angled triangle where the sides containing the right angle are $15 \cdot 8$ m and $8 \cdot 9$ m.*

Fig. 56 illustrates the problem.

To find C, $\tan C = \dfrac{8 \cdot 9}{15 \cdot 8} = 0 \cdot 5633 = \tan 29° 24'$.

To find A, $\tan A = \dfrac{15 \cdot 8}{8 \cdot 9} = 1 \cdot 7753 = \tan 60° 36'$.

These should be checked by seeing if their sum is 90°.

To find AC.

(1) $AC = \sqrt{15\cdot8^2 + 8\cdot9^2} = 18\cdot1$ m approx., or

(2) $$\frac{AC}{8\cdot9} = \text{cosec } C$$

$\therefore \quad AC = 8\cdot9 \text{ cosec } C$

$\begin{aligned}
\log AC &= \log 8\cdot9 + \log \text{cosec } C \\
&= 0\cdot9494 + 0\cdot3090 \\
&= 1\cdot2584 \\
&= \log 18\cdot13
\end{aligned}$

$\therefore \quad AC = 18\cdot1$ m approx.

(b) Given one angle and the hypotenuse.

Example 2. *Solve the right-angled triangle in which one angle is 27° 43' and the hypotenuse is 6·85 cm.*

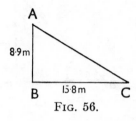

FIG. 56.

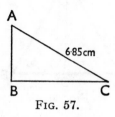

FIG. 57.

In Fig. 57 $\qquad C = 27° \ 43'$

$\therefore \quad A = 90° - C = 90 - 27° \ 43'$

$\qquad\qquad = 62° \ 17'.$

To find AB and BC

$\begin{aligned}
AB &= AC \sin ACB \\
&= 6\cdot85 \times \sin 27° \ 43' \\
&= 3\cdot19 \text{ cm.} \\
BC &= AC \cos ACB \\
&= 6\cdot85 \times \cos 27° \ 43' \\
&= 6\cdot06 \text{ cm.}
\end{aligned}$

These examples will serve to indicate the methods to be adopted in other cases.

(c) **Special cases.**

(1) **The equilateral triangle.**

In Fig. 58 ABC is an equilateral triangle, AD is the perpendicular bisector of the base.

It also bisects $\angle CAB$ (Theorem **3**, § 11).

$\therefore \quad \angle DAB = 30°$

and $\qquad\qquad \angle ABD = 60°$

Let each side of the $\triangle$ be a units of length.

Then $$DB = \frac{a}{2}$$

$$\therefore \quad AD = \sqrt{AB^2 - DB^2} \qquad \text{(Theorem 9)}$$

$$= \sqrt{a^2 - \frac{a^2}{4}}$$

$$= \sqrt{\frac{3a^2}{4}}$$

$$= a \times \frac{\sqrt{3}}{2}$$

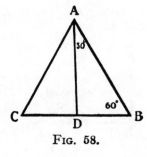

FIG. 58.

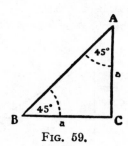

FIG. 59.

$$\therefore \quad \sin 60° = \frac{AD}{AB} = \frac{a \times \frac{\sqrt{3}}{2}}{a} = \frac{\sqrt{3}}{2}$$

$$\cos 60° = \frac{BD}{AB} = \frac{a}{2} \div a = \frac{1}{2}$$

$$\tan 60° = \frac{AD}{DB} = \frac{a\sqrt{3}}{2} \div \frac{a}{2} = \sqrt{3}$$

Similarly

$$\sin 30° = \frac{BD}{AB} = \frac{a}{2} \div a = \frac{1}{2}$$

$$\cos 30° = \frac{AD}{AB} = \frac{a\sqrt{3}}{2} \div a = \frac{\sqrt{3}}{2}$$

$$\tan 30° = \frac{DB}{AD} = \frac{a}{2} \div \frac{a\sqrt{3}}{2} = \frac{1}{\sqrt{3}}$$

Note.—The ratios for 30° can be found from those for 60° by using the results of §§ 48 and 53.

(2) **The right-angled isosceles triangle.**

Fig. 59 represents an isosceles triangle with $AC = BC$ and $\angle ACB = 90°$.

Let each of the equal sides be a units of length.

Then
$$AB^2 = AC^2 + BC^2 \qquad \text{(Theorem 9)}$$
$$= a^2 + a^2$$
$$= 2a^2$$
$$\therefore \quad AB = a\sqrt{2}$$

$$\therefore \quad \sin 45° = \frac{AC}{AB} = \frac{a}{a\sqrt{2}} = \frac{1}{\sqrt{2}}$$

$$\cos 45° = \frac{BC}{AB} = \frac{a}{a\sqrt{2}} = \frac{1}{\sqrt{2}}$$

$$\tan 45° = \frac{AC}{BC} = \frac{a}{a} = 1$$

It should be noted that $\triangle ABC$ represents half a square of which AB is the diagonal.

62. Slope and gradient.

Fig. 60 represents a side view of the section of a path AC in which AB represents the horizontal level and BC the vertical rise.

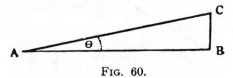

FIG. 60.

$\angle CAB$, denoted by θ, is the angle between the plane of the path and the horizontal.

Then $\angle CAB$ is called the *angle of slope of the path* or more briefly $\angle CAB$ *is the slope of the path*.

Now
$$\tan \theta = \frac{CB}{AB}$$

This tangent is called the gradient of the path.
Generally, if θ be the slope of a path, $\tan \theta$ is the gradient.
A gradient is frequently given in the form 1 in 55, and in this form can be seen by the side of railways to denote the gradient of the rails. This means that the *tangent of the angle of slope* is $\frac{1}{55}$.
When the angle of slope is very small, as happens in the case of a railway and most roads, it makes little practical difference if instead of the tangent $\left(\frac{CB}{AB}\text{ in Fig. 60}\right)$ we take $\frac{CB}{AC}$, *i.e.* the sine of the angle instead of the tangent. In

practice also it is easier to measure AC, and the difference between this and AB is relatively small, provided the angle is small.

If the student refers to the tables of tangents and sines he will see how small is the differences between them for small angles.

63. Projections.

In Chapter I, § 22, we referred to the projection of a straight line on a plane.   We will now examine this further.

Projection of a straight line on a fixed line.

In Fig. 61, let PQ be a straight line of unlimited length, and AB another straight line which, when produced to meet PQ at O, makes an angle θ with it.

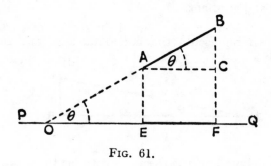

Fig. 61.

From A and B draw perpendiculars to meet PQ at E and F.

Draw AC parallel to EF.

EF is called the projection of AB on PQ (§ 22).

Now $\angle BAC = \angle BOF = \theta$ (Theorem 2)
and $EF = AC$
Also $AC = AB \cos \theta$ (§ 47)
          $\therefore$ **$EF = AB \cos \theta$**.

$\therefore$ *If a straight line AB, produced if necessary, makes an angle θ with another straight line, the length of its projection on that straight line is AB cos θ.*

It should be noted in Fig. 61 that

$$BC = AB \sin \theta$$

From which it is evident that if we draw a straight line at right angles to PQ, the *projection of AB upon such a straight line is AB sin θ.*

Exercise 10

General questions on the trigonometrical ratios.

1. In a right-angled triangle the two sides containing the right angle are 2·34 m and 1·64 m. Find the angles and the hypotenuse.

2. In a triangle ABC, C being a right angle, AC is 122 cm, AB is 175 cm. Compute the angle B.

3. In a triangle ABC, $C = 90°$. If $A = 37° 21'$ and $c = 91·4$, find a and b.

4. ABC is a triangle, the angle C being a right angle. AC is 21·32 m, BC is 12·56 m. Find the angles A and B.

5. In a triangle ABC, AD is the perpendicular on BC: AB is 3·25 cm, B is 55°, BC is 4·68 cm. Find the length of AD. Find also BD, DC and AC.

6. ABC is a right-angled triangle, C being the right angle. If $a = 378$ mm and $c = 543$ mm, find A and b.

7. A ladder 20 m long rests against a vertical wall. By means of trigonometrical tables find the inclination of the ladder to the horizontal when the foot of the ladder is:
   (1) 7 m from the wall.
   (2) 10 m from the wall.

8. A ship starts from a point O and travels 18 km h^{-1} in a direction 35° north of east. How far will it be north and east of O after an hour?

9. A pendulum of length 20 cm swings on either side of the vertical through an angle of 15°. Through what height does the bob rise?

10. If the side of an equilateral triangle is x m, find the altitude of the triangle. Hence find sin 60° and sin 30°.

11. Two straight lines OX and OY are at right angles to one another. A straight line 3·5 cm long makes an angle of 42° with OX. Find the lengths of its projections on OX and OY.

12. A man walking 1·5 km up the line of greatest slope of a hill rises 94 m. Find the gradient of the hill.

13. A ship starts from a given point and sails 15·5 km in a direction 41° 15' west of north. How far has it gone west and north respectively?

14. A point P is 14·5 km north of Q and Q is 9 km west of R. Find the bearing of P from R and its distance from R.

CHAPTER IV

RELATIONS BETWEEN THE TRIGONOMETRICAL RATIOS

64. Since each of the trigonometrical ratios involves two of the three sides of a right-angled triangle, it is to be expected that definite relations exist between them. These relations are very important and will constantly be used in further work. The most important of them will be proved in this chapter.

$$\tan \theta = \frac{\sin \theta}{\cos \theta}.$$

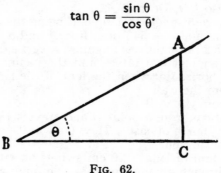

FIG. 62.

Let ABC (Fig. 62) be any acute angle (θ). From a point A on one arm draw AC perpendicular to the other arm.

Then

$$\sin \theta = \frac{AC}{AB}$$

and

$$\cos \theta = \frac{BC}{AB}$$

$$\therefore \quad \frac{\sin \theta}{\cos \theta} = \frac{AC}{AB} \div \frac{BC}{AB}$$

$$= \frac{AC}{AB} \times \frac{AB}{BC}$$

$$= \frac{AC}{BC}$$

$$= \tan \theta$$

$$\therefore \quad \frac{\sin \theta}{\cos \theta} = \tan \theta. \tag{1}$$

Similarly we may prove that $\cot \theta = \dfrac{\cos \theta}{\sin \theta}$

65. $\sin^2 \theta + \cos^2 \theta = 1$.

From Fig. 62

$$AC^2 + BC^2 = AB^2 \quad \text{(Theorem of Pythagoras, § 14)}$$

Dividing throughout by AB^2

we get
$$\frac{AC^2}{AB^2} + \frac{BC^2}{AB^2} = 1$$

$$\therefore \quad (\sin \theta)^2 + (\cos \theta)^2 = 1$$

or as usually written

$$\sin^2 \theta + \cos^2 \theta = 1 \tag{2}$$

This very important result may be transformed and used to find either of the ratios when the other is given.

Thus
$$\sin^2 \theta = 1 - \cos^2 \theta$$

$$\therefore \quad \sin \theta = \sqrt{1 - \cos^2 \theta}$$

Similarly
$$\cos \theta = \sqrt{1 - \sin^2 \theta}$$

Combining formulae (1) and (2)

$$\tan \theta = \frac{\sin \theta}{\cos \theta}$$

becomes
$$\tan \theta = \frac{\sin \theta}{\sqrt{1 - \sin^2 \theta}}$$

This form expresses the *tangent* in terms of the *sine* only. It may similarly be expressed in terms of the cosine

thus
$$\tan \theta = \frac{\sqrt{1 - \cos^2 \theta}}{\cos \theta}$$

66.
$$1 + \tan^2 \theta = \sec^2 \theta$$
$$1 + \cot^2 \theta = \operatorname{cosec}^2 \theta$$

Using the formula $\quad \sin^2 \theta + \cos^2 \theta = 1$
and dividing throughout by $\cos^2 \theta$

we get
$$\frac{\sin^2 \theta}{\cos^2 \theta} + 1 = \frac{1}{\cos^2 \theta}$$

$$\therefore \quad \tan^2 \theta + 1 = \sec^2 \theta$$

Again, dividing throughout by $\sin^2 \theta$

we get
$$1 + \frac{\cos^2 \theta}{\sin^2 \theta} = \frac{1}{\sin^2 \theta}$$

$$\therefore \quad 1 + \cot^2 \theta = \operatorname{cosec}^2 \theta.$$

We may also write these formulae in the forms

$$\tan^2 \theta = \sec^2 \theta - 1$$

and
$$\cot^2 \theta = \operatorname{cosec}^2 \theta - 1.$$

Using these forms we can change tangents into secants and cotangents into cosecants and vice versa when it is necessary in a given problem.

Exercise 11

1. Find $\tan \theta$ when $\sin \theta = 0 \cdot 5736$ and $\cos \theta = 0 \cdot 8192$.
2. If $\sin \theta = \frac{2}{3}$, find $\cos \theta$ and $\tan \theta$.
3. Find $\sin \theta$ when $\cos \theta = 0 \cdot 47$.
4. Find $\sec \theta$ when $\tan \theta = 1 \cdot 2799$.
5. If $\sec \theta = 1 \cdot 2062$ find $\tan \theta$, $\cos \theta$ and $\sin \theta$.
6. Find $\operatorname{cosec} \theta$ when $\cot \theta = 0 \cdot 5774$.
7. If $\cot \theta = 1 \cdot 63$, find $\operatorname{cosec} \theta$, $\sin \theta$ and $\cos \theta$.
8. If $\tan \theta = t$, find expressions for $\sec \theta$, $\cos \theta$ and $\sin \theta$ in terms of t.
9. If $\cos \alpha = 0 \cdot 4695$, find $\sin \alpha$ and $\tan \alpha$.
10. Prove that $\tan \theta + \cot \theta = \sec \theta \operatorname{cosec} \theta$.

CHAPTER V

TRIGONOMETRICAL RATIOS OF ANGLES IN THE SECOND QUADRANT

67. IN Chapter III we dealt with the trigonometrical ratios of *acute* angles, or angles in the first quadrant. It will be remembered that in Chapter I, § 5, when considering the meaning of an angle as being formed by the rotation of a straight line from a fixed position, we saw that there was no limit to the amount of rotation and consequently angles could be of any magnitude.

We must now consider the extension of trigonometrical ratios to angles greater than a right angle. At the present, however, we shall not examine the general question of angles of any magnitude, but confine ourselves to *obtuse* angles, or angles in the second quadrant, as these are necessary in many practical applications of trigonometry.

68. Positive and negative lines.

Before proceeding to deal with the trigonometrical ratios of obtuse angles it is necessary to consider the methods by which we distinguish between measurements made on a straight line in opposite directions. These will be familiar to those who have studied co-ordinates and graphs. It is desirable, however, to revise the principles involved before applying them to trigonometry.

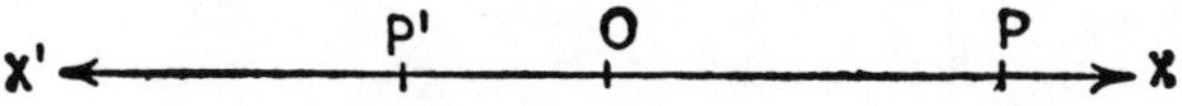

FIG. 63.

Let Fig. 63 represent a straight road XOX'.

If now a man travels 4 miles from O to P in the direction $\overrightarrow{OX}$ and then turns and travels 6 miles in the opposite direction to P', the net result is that he has travelled $(4 - 6)$ miles, i.e. -2 miles from O. The significance of the negative sign is that the man is now 2 *miles in the opposite direction* from that in which he started.

In such a way as this we arrive at the convention by which we agree to use $+$ and $-$ signs to indicate opposite directions.

If now we consider two straight lines at right angles to one another, as $X'OX$, $Y'OY$, in Fig. 64, such as are used for co-ordinates and graphs, we can extend to these the con-

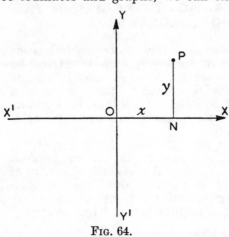

FIG. 64.

ventions used for one straight line as indicated above. The lines OX, OY are called the *axes of co-ordinates*. OX measures the x-co-ordinate, called the *abscissa*, and OY measures the y-co-ordinate, called the *ordinate*. Any point P (Fig. 64), has a *pair* of co-ordinates (x, y). Each pair determines a unique point.

The area of the diagram, Fig. 65, is considered to be divided into four quadrants as shown. Values of x measured to the right are

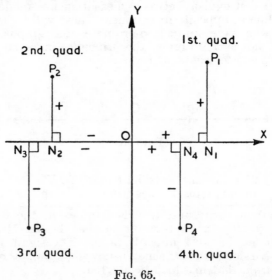

FIG. 65.

$+ ve$, and to the left are $- ve$. Values of y measured upwards are $+ ve$, and downwards are $- ve$. This is a universally accepted convention.

P_1 lies in the first quadrant and N_1 is the foot of the perpendicular from P_1 to OX. $\overrightarrow{ON_1}$ is in the direction of $\overrightarrow{OX}$ and is $+ ve$; $\overrightarrow{N_1P_1}$ is in the direction of $\overrightarrow{OY}$ and is $+$ ve. Thus the co-ordinates of any point P_1 in the first quadrant are $(+, +)$.

P_2 lies in the second quadrant and N_2 is the foot of the perpendicular from P_2 to OX. $\overrightarrow{ON_2}$ is in the direction of $\overrightarrow{XO}$ and is $- ve$; $\overrightarrow{N_2P_2}$ is in the direction of $\overrightarrow{OY}$ and is $+ ve$. Thus the co-ordinates of any point P_2 in the second quadrant are $(-, +)$.

Similarly the co-ordinates of P_3 in the third quadrant are $(-, -)$, and of P_4 in the fourth quadrant are $(+, -)$.

At present we shall content ourselves with considering *points in the first two quadrants*. The general problem for all four quadrants is discussed later (Chapter XI).

69. Direction of Rotation of Angle.

The *direction in which the rotating line turns* must be taken into account when considering the angle itself.

Thus in Fig. 66 the angle $A\hat{O}B$ may be formed by rotation

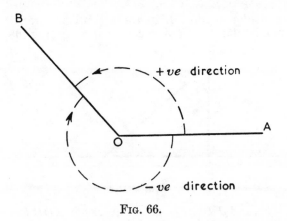

Fig. 66.

in an anti-clockwise direction or by rotation in a clockwise direction.

By convention an *anti-clockwise* rotation is *positive* and a *clockwise* rotation is *negative*.

Negative angles will be considered further in Chapter XI. In the meantime, we shall use positive angles formed by anti-clockwise rotation.

70. The Sign Convention for the Hypotenuse.

Consider a point A in the first quadrant. Draw AD perpendicular to $X'OX$ meeting it at D (Fig. 67).

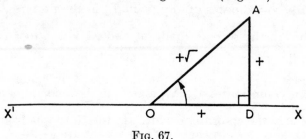

Fig. 67.

OD is $+ve$ and DA is $+ve$. The angle XOA = angle DOA, which is acute.

Also
$$OA^2 = OD^2 + DA^2$$
$$= (+ve)^2 + (+ve)^2 = +ve \text{ quantity}$$
$$= a^2 \text{ (say where } a \text{ is } +ve)$$

Now the equation $OA^2 = a^2$ has two roots $OA = a$ or $OA = -a$, so we must decide on a sign convention. *We take OA as the $+ve$ root.*

Now consider a point B in the second quadrant. Draw BE perpendicular to $X'OX$ meeting it at E (Fig. 68).

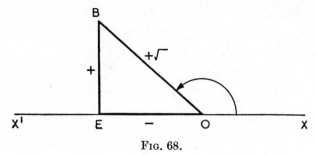

Fig. 68.

OE is $-ve$ and EB is $+ve$. The angle XOB (= $180°$ − angle EOB) is obtuse.

Also
$$OB^2 = OE^2 + EB^2$$
$$= (-ve)^2 + (+ve)^2$$
$$= (+ve) + (+ve) = +ve \text{ quantity.}$$

We have already decided on a sign convention for the root, so OB is $+ ve$.

Now the sides required to give the ratios of $\angle XOB$ are the same as those needed for its supplement $\angle EOB$. The only change which may have taken place is in the *sign* prefixed to the length of a side. OD ($+ ve$ in Fig. 67) has become OE ($- ve$ in Fig. 68).

Thus we have the following rules:

RATIO	ACUTE ANGLE	OBTUSE ANGLE
SIN	+	+
COS	+	−
TAN	+	−

Fig. 69.

We see this at once by combining Fig. 67 and Fig. 68 into Fig. 70.

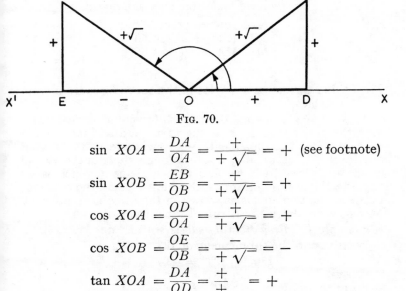

Fig. 70.

$$\sin XOA = \frac{DA}{OA} = \frac{+}{+\sqrt{}} = + \text{ (see footnote)}$$

$$\sin XOB = \frac{EB}{OB} = \frac{+}{+\sqrt{}} = +$$

$$\cos XOA = \frac{OD}{OA} = \frac{+}{+\sqrt{}} = +$$

$$\cos XOB = \frac{OE}{OB} = \frac{-}{+\sqrt{}} = -$$

$$\tan XOA = \frac{DA}{OD} = \frac{+}{+} = +$$

$$\tan XOB = \frac{EB}{OE} = \frac{+}{-} = -$$

Note.—We use here the abbreviations $+$ and $-$ to stand for *a positive quantity* and *a negative quantity* respectively.

Further, by making $\triangle OBE \equiv \triangle OAD$ in Fig. 70 and using the rules we see that

sine of an angle   = sine of its supplement
cosine of an angle = − cosine of its supplement
tangent of an angle = − tangent of its supplement.

These results may alternatively be expressed thus:

$$\sin \theta = \sin (180° − \theta)$$
$$\cos \theta = − \cos (180° − \theta)$$
$$\tan \theta = − \tan (180° − \theta).$$

e.g.
$$\left. \begin{array}{l} \sin 100° = \sin 80° \\ \cos 117° = − \cos 63° \\ \tan 147° = − \tan 33° \end{array} \right\}$$

The reciprocal ratios, cosecant, secant and cotangent will have the same signs as the ratios from which they are derived.

∴   *cosecant* has same sign as *sine*
     secant has same sign as *cosine*
     cotangent has same sign as *tangent.*

e.g.
$$\cosec 108° = \cosec 72°$$
$$\sec\ \ \ 121° = − \sec 59°$$
$$\cot\ \ \ 154° = − \cot 36°$$

71. To find the ratios of angles in the second quadrant from the tables.

As will have been seen, the tables of trigonometrical ratios give the ratios of angles in the first quadrant only. But each of these is supplementary to an angle in the second quadrant. Consequently if a ratio of an angle in the second quadrant is required, we find its supplement which is an angle in the first quadrant, and then, by using the relations between the two angles as shown in the previous paragraph we can write down the required ratio from the tables.

Example I. *Find from the tables sin 137° and cos 137°.*

We first find the supplement of 137° which is
$$180° − 137° = 43°.$$
∴ by § 70 $\sin 137° = \sin 43°.$
From the tables $\sin 43° = 0{\cdot}6820$
∴ $\sin 137° = 0{\cdot}6820.$
Again $\cos \theta = − \cos (180° − \theta)$
∴ $\cos 137° = − \cos (180° − 137°)$
      $= − \cos 43°$
      $= − 0{\cdot}7314.$

Example 2. *Find the values of tan 162° and sec 162°.*

From the above $\quad \tan \theta = - \tan (180° - \theta)$

$$\therefore \quad \tan 162° = - \tan (180° - 162°)$$
$$= - \tan 18°$$
$$= - 0.3249.$$

Also $\qquad\qquad \sec \theta = - \sec (180° - \theta)$

$$\therefore \quad \sec 162° = - \sec (180° - 162°)$$
$$= - \sec 18°$$
$$= - 1.0515.$$

72. Ratios for 180°.

These can be found either by using the same arguments as were employed in the cases of 0° and 90° or by applying the above relation between an angle and its supplement.

From these we conclude

$$\sin 180° = 0$$
$$\cos 180° = - 1$$
$$\tan 180° = 0.$$

73. To find an angle when a ratio is given.

When this converse problem has to be solved in cases where the angle may be in the second quadrant, difficulties arise which did not occur when dealing with angles in the first quadrant only. The following examples will illustrate these.

Example 1. *Find the angle whose cosine is — 0.5577.*

The negative sign for a cosine shows that the angle is in the second quadrant, since $\cos \theta = - \cos (180° - \theta)$.

From the tables we find that

$$\cos 56° 6' = + 0.5577.$$

$\therefore$ the angle required is the supplement of this

$$i.e. \quad 180° - 56° 6'$$
$$= 123° 54'.$$

Example 2. *Find the angles whose sine + 0.9483.*

We know that *since an angle and its supplement have the same sine*, there are two angles with the sine + 0.9483, and they are supplementary.

From the tables $\quad \sin 71° 30' = + 0.9483.$

$\therefore$ Since $\qquad\qquad \sin \theta = \sin (180° - \theta)$

$$\therefore \quad \sin 71° 30' = \sin (180° - 71° 30')$$
$$= \sin 108° 30'.$$

There are therefore two answers, 71° 30' and 108° 30', and *there are always two angles having a given sine*, one in the first and one in the second quadrant. Which of these is the angle required when solving some problem must be determined by the special conditions of the problem.

Example 3. *Find the angle whose tangent is* — *1·3764.*

Since the tangent is negative, the angle required must lie in the second quadrant.

From the tables

$$\tan 54° = + 1·3764$$

and since
$$\tan \theta = - \tan (180° - \theta)$$
$$\therefore - 1·3764 = \tan (180° - 54°)$$
$$= \tan 126°.$$

74. Inverse notation.

The sign " $tan^{-1} - 1·3674$ " is employed to signify " *the angle whose tangent is* — *1·3674* "

And, in general

> $sin^{-1} x$ *means* " *the angle whose sine is* x "
> $cos^{-1} x$ *means* " *the angle whose cosine is* x ",

etc.

Three points should be noted.

(1) $\sin^{-1} x$ stands for an *angle* : thus $\sin^{-1} \frac{1}{2} = 30°$.

(2) The " $- 1$ " is not an index, but merely a sign to denote inverse notation.

(3) $(\sin x)^{-1}$ is not used, because by § 31 it would mean the reciprocal of $\sin x$ and this is $\operatorname{cosec} x$.

75. Ratios of some important angles.

We are now able to tabulate the values of the sine, cosine and tangents of certain angles between 0° and 180°. The table will also state in a convenient form the ratios of a few important angles. They should be memorised.

	0°.	30°.	45°.	60°.	90°.	120°.	135°.	150°.	180°.
Sine .	0	*Increasing and Positive.*				*Decreasing and Positive.*			0
	0	$\frac{1}{2}$	$\frac{1}{\sqrt{2}}$	$\frac{\sqrt{3}}{2}$	1	$\frac{\sqrt{3}}{2}$	$\frac{1}{\sqrt{2}}$	$\frac{1}{2}$	0
Cosine .		*Decreasing and Positive.*				*Decreasing and Negative.*			
	1	$\frac{\sqrt{3}}{2}$	$\frac{1}{\sqrt{2}}$	$\frac{1}{2}$	0	$-\frac{1}{2}$	$-\frac{1}{\sqrt{2}}$	$-\frac{\sqrt{3}}{2}$	-1
Tangent		*Increasing and Positive.*				*Increasing and Negative.*			
	0	$\frac{1}{\sqrt{3}}$	1	$\sqrt{3}$	∞	$-\sqrt{3}$	-1	$-\frac{1}{\sqrt{3}}$	0

76. Graphs of sine, cosine and tangent between 0° and 180°.

The changes in the ratios of angles in the first and second quadrants are made clear by drawing their graphs. This

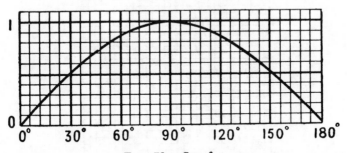

FIG. 71. SIN θ.

may be done by using the values given in the above table or, more accurately, by taking values from the tables.

An inspection of these graphs will illustrate the results reached in § 73 (second example).

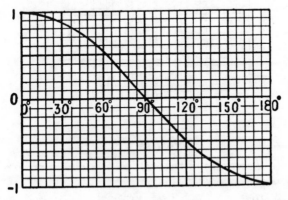

FIG. 72. COS θ.

It is evident from Fig. 71, that there are two angles, one in each quadrant with a given sine.

From Figs. 72 and 73, it will be seen that there is only one angle between 0° and 180° corresponding to a given cosine or tangent.

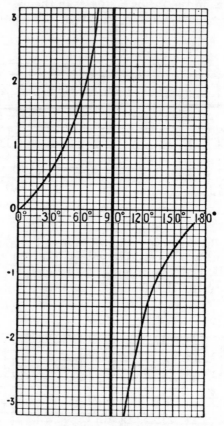

FIG. 73.   TAN θ.

Exercise 12

1. Write down from the tables the sines, cosines and tangents of the following angles:

    (a) 102°.   (b) 149° 33′.   (c) 109° 28′.
    (d) 145° 16′.   (e) 154° 36′.

2. Find θ when:

    (a) $\sin \theta = 0 \cdot 6508$.   (b) $\sin \theta = 0 \cdot 9126$.
    (c) $\sin \theta = 0 \cdot 3469$.   (d) $\sin \theta = 0 \cdot 7122$.

3. Find the angles whose cosines are:

    (a) -0.4540.   (b) -0.8131.   (c) -0.1788.
    (d) -0.9354.   (e) -0.7917.   (f) -0.9154.

4. Find θ when:

    (a) $\tan \theta = -0.5543$.   (b) $\tan \theta = -1.4938$.
    (c) $\tan \theta = -2.4383$.   (d) $\tan \theta = -1.7603$.
    (e) $\tan \theta = -0.7142$.   (f) $\tan \theta = -1.1757$.

5. Find the values of:

    (a) cosec $154°$.   (b) sec $162° 30'$.
    (c) cot $163° 12'$.

6. Find θ when:

    (a) $\sec \theta = -1.6514$.   (b) $\sec \theta = -2.1301$.
    (c) $\operatorname{cosec} \theta = 1.7305$.   (d) $\operatorname{cosec} \theta = 2.4586$.
    (e) $\cot \theta = -1.6643$.   (f) $\cot \theta = -0.3819$.

7. Find the value of $\dfrac{\tan A}{\sec B}$ when $A = 150°$, $B = 163° 17'$.

8. Find the values of:

    (a) $\sin^{-1} 0.9336$.   (b) $\cos^{-1} 0.4226$.
    (d) $\tan^{-1} 1.3764$.   (d) $\cos^{-1} -0.3907$.

CHAPTER VI

TRIGONOMETRICAL RATIOS OF COMPOUND ANGLES

77. WE often need to use the trigonometrical ratios of the sum or difference of two angles. If A and B are any two angles, $(A + B)$ and $(A - B)$ are usually called compound angles, and it is convenient to be able to express their trigonometrical ratios in terms of the ratios of A and B.

The beginner must beware of thinking that $\sin (A + B)$ is equal to $(\sin A + \sin B)$. He should test this by taking the values of $\sin A$, $\sin B$, and $\sin (A + B)$ for some particular values of A and B from the tables and comparing them.

78. We will first show that:

$$\sin (A + B) = \sin A \cos B + \cos A \sin B$$
and $$\cos (A + B) = \cos A \cos B - \sin A \sin B$$

To simplify the proof at this stage we will assume that A, B, and $(A + B)$ are all acute angles.

The student is advised to make his own diagram step by step with the following construction.

Construction.

Let a staight line rotating from a position on a fixed line OX trace out (1) the angle XOY, equal to A and YOZ equal to B (Fig. 74).

Then $$\angle XOZ = (A + B)$$

In OZ take any point P.

Draw PQ perpendicular to OX and PM perpendicular to OY.

From M draw MN perpendicular to OX and MR parallel to OX.

Then $$MR = QN$$
Proof

$$\angle RPM = 90° - \angle PMR$$
$$= \angle RMO$$
But $$\angle RMO = \angle MOX \qquad \text{(Theorem 2, § 9)}$$
$$= A$$
$$\therefore \angle RPM = A$$

98

Again sin $(A + B) = \sin XOZ$

$$= \frac{PQ}{OP}$$

$$= \frac{RQ + PR}{OP}$$

$$= \frac{RQ}{OP} + \frac{PR}{OP}$$

$$= \frac{MN}{OP} + \frac{PR}{OP}$$

$$= \left(\frac{MN}{OM} \times \frac{OM}{OP}\right) + \left(\frac{PR}{PM} \times \frac{PM}{OP}\right)$$

$$= \sin A \cos B + \cos A \sin B.$$

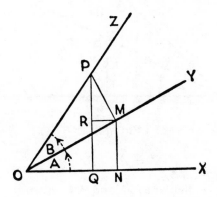

Fig. 74.

Note the device of introducing $\frac{OM}{OM}$ and $\frac{PM}{PM}$, each of which is unity, into the last line but one.

Again

$\cos (A + B) = \cos XOZ$

$$= \frac{OQ}{OP}$$

$$= \frac{ON - NQ}{OP}$$

$$= \frac{ON}{OP} - \frac{NQ}{OP}$$

$$= \frac{ON}{OP} - \frac{RM}{OP}$$

$$= \left(\frac{ON}{OM} \times \frac{OM}{OP}\right) - \left(\frac{RM}{PM} \times \frac{PM}{OP}\right)$$

$$= \cos A \cos B - \sin A \sin B.$$

79. We will next prove the corresponding formulae for $(A - B)$, viz.:

$$\sin (A - B) = \sin A \cos B - \cos A \sin B$$
$$\cos (A - B) = \cos A \cos B + \sin A \sin B$$

Construction.

Let a straight line rotating from a fixed position on OX describe an angle XOY, equal to A, and then, rotating in an opposite direction, describe an angle YOZ, equal to B (Fig. 75).

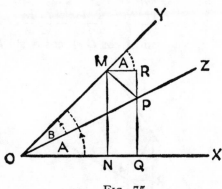

FIG. 75.

Then $\qquad XOZ = A - B.$

Take a point P on OZ.

Draw PQ perpendicular to OX and PM perpendicular to OY. From M draw MN perpendicular to OX and MR parallel to OX to meet PQ *produced* in R.

Proof
$$\angle RPM = 90° - \angle PMR$$
$$= \angle RMY \quad \text{(since } PM \text{ is perp. to } OY)$$
$$= \angle YOX \qquad \text{(Theorem 2, § 9)}$$
Now $\qquad\qquad\quad = A.$
$$\sin (A - B) = \sin XOZ$$
$$= \frac{PQ}{OP}$$
$$= \frac{RQ - RP}{OP}$$
$$= \frac{RQ}{OP} - \frac{RP}{OP}$$
$$= \frac{MN}{OP} - \frac{RP}{OP}$$
$$= \left(\frac{MN}{OM} \times \frac{OM}{OP}\right) - \left(\frac{RP}{PM} \times \frac{PM}{OP}\right)$$
$$= \sin A \cos B - \cos A \sin B.$$

Again

$$\begin{aligned}
\cos (A - B) &= \cos XOZ \\
&= \frac{OQ}{OP} \\
&= \frac{ON + QN}{OP} \\
&= \frac{ON}{OP} + \frac{QN}{OP} \\
&= \frac{ON}{OP} + \frac{RM}{OP} \\
&= \left(\frac{ON}{OM} \times \frac{OM}{OP}\right) + \left(\frac{RM}{PM} \times \frac{PM}{OP}\right) \\
&= \cos A \cos B + \sin A \sin B.
\end{aligned}$$

80. These formulae have been proved for acute angles only, but they can be shown to be true for angles of any size. They are of great importance. We collect them for reference:

$$\sin (A + B) = \sin A \cos B + \cos A \sin B \qquad (1)$$
$$\cos (A + B) = \cos A \cos B - \sin A \sin B \qquad (2)$$
$$\sin (A - B) = \sin A \cos B - \cos A \sin B \qquad (3)$$
$$\cos (A - B) = \cos A \cos B + \sin A \sin B \qquad (4)$$

81. From the above we may find similar formulae for $\tan (A + B)$ and $\tan (A - B)$ as follows:

$$\begin{aligned}
\tan (A + B) &= \frac{\sin (A + B)}{\cos (A + B)} \\
&= \frac{\sin A \cos B + \cos A \sin B}{\cos A \cos B - \sin A \sin B}
\end{aligned}$$

Dividing numerator and denominator by $\cos A \cos B$

we get $\quad \tan (A + B) = \dfrac{\dfrac{\sin A \cos B}{\cos A \cos B} + \dfrac{\cos A \sin B}{\cos A \cos B}}{\dfrac{\cos A \cos B}{\cos A \cos B} - \dfrac{\sin A \sin B}{\cos A \cos B}}$

$$= \frac{\dfrac{\sin A}{\cos A} + \dfrac{\sin B}{\cos B}}{1 - \dfrac{\sin A}{\cos A} \cdot \dfrac{\sin B}{\cos B}}$$

$$\therefore \quad \tan (A + B) = \frac{\tan A + \tan B}{1 - \tan A \tan B}$$

Similarly we may show

$$\tan (A - B) = \frac{\tan A - \tan B}{1 + \tan A \tan B}$$

with similar formulae for cotangents.

82. Worked Examples.

Example I. *Using the values of the sines and cosines of 30° and 45° as given in the table in § 75, find sin 75°.*

Using
$$\sin (A + B) = \sin A \cos B + \cos A \sin B$$
and substituting
$$A = 45°, B = 30°$$
we have $\sin 75° = \sin 45° \cos 30° + \cos 45° \sin 30°$

$$= \left(\frac{1}{\sqrt{2}} \times \frac{\sqrt{3}}{2} \right) + \left(\frac{1}{\sqrt{2}} \times \frac{1}{2} \right)$$

$$= \frac{\sqrt{3}}{2\sqrt{2}} + \frac{1}{2\sqrt{2}}$$

$$= \frac{\sqrt{3} + 1}{2\sqrt{2}}.$$

Example 2. *If cos a = 0·6 and cos β = 0·8, find the values of sin (a + β) and cos (a + β), without using the tables.*

We must first find sin a and sin β. For these we use the results given in § 65.

$$\sin a = \sqrt{1 - \cos^2 a}$$

Substituting the given value of cos a

$$\sin a = \sqrt{1 - (0·6)^2}$$
$$= \sqrt{1 - 0·36}$$
$$= \sqrt{0·64}$$
$$= 0·8.$$

Similarly we find sin β = 0·6.

Using $\sin (A + B) = \sin A \cos B + \cos A \sin B$
and substituting we have

$$\sin (a + β) = (0·8 \times 0·8) + (0·6 \times 0·6)$$
$$= 0·64 + 0·36$$
$$= 1$$

Also $\cos (a + β) = \cos a \cos β - \sin a \sin β$
$$= (0·6 \times 0·8) - (0·8 \times 0·6)$$
$$= 0.$$

Obviously   a + β = 90°, since cos 90° = 0.
∴ a and β are complementary.

Exercise 13

1. If $\cos A = 0.2$ and $\cos B = 0.5$, find the values of $\sin (A + B)$ and $\cos (A - B)$.

2. Use the ratios of $45°$ and $30°$ from the table in § 75 to find the values of $\sin 15°$ and $\cos 75°$.

3. By using the formula for $\sin (A - B)$ prove that:
$$\sin (90° - \theta) = \cos \theta.$$

4. By means of the formulae of § 80, find $\sin (A - B)$ when $\sin B = 0.23$ and $\cos A = 0.309$.

5. Find $\sin (A + B)$ and $\tan (A + B)$ when $\sin A = 0.71$ and $\cos B = 0.32$.

6. Use the formula of $\tan (A + B)$ to find $\tan 75°$.

7. Find $\tan (A + B)$ and $\tan (A - B)$ when $\tan A = 1.2$ and $\tan B = 0.4$.

8. By using the formula for $\tan (A - B)$ prove that
$$\tan (180° - A) = - \tan A.$$

9. Find the values of:
    (1) $\sin 52° \cos 18° - \cos 52° \sin 18°$.
    (2) $\cos 73° \cos 12° + \sin 73° \sin 12°$.

10. Find the values of: (a) $\dfrac{\tan 52° + \tan 16°}{1 - \tan 52° \tan 16°}$.

                               (b) $\dfrac{\tan 64° - \tan 25°}{1 + \tan 64° \tan 25°}$.

11. Prove that $\sin (\theta + 45°) = \dfrac{1}{\sqrt{2}} (\sin \theta + \cos \theta)$.

12. Prove that $\tan (\theta + 45°) = \dfrac{\tan \theta + 1}{1 - \tan \theta}$.

83. Multiple and sub-multiple angle formulae.

From the preceding formulae we may deduce others of great practical importance.

From § 78 $\sin (A + B) = \sin A \cos B + \cos A \sin B$.

There have been no limitations of the angles.

$\therefore$ let $B = A$.

Substituting

$$\sin 2A = \sin A \cos A + \cos A \sin A$$

or $\sin 2A = 2 \sin A \cos A$ (1)

If $2A$ be replaced by θ

then $\sin \theta = 2 \sin \dfrac{\theta}{2} \cos \dfrac{\theta}{2}$ (2)

We may use whichever of these formulae is more convenient in a given problem.

Again $\qquad \cos(A + B) = \cos A \cos B - \sin A \sin B$

Let $\qquad\qquad B = A,$

then $\qquad\qquad \cos 2A = \cos^2 A - \sin^2 A \qquad\qquad$ (4)

This may be transformed into formulae giving $\cos A$ or $\sin^2 A$ in terms of ^2A.

Since $\quad \sin^2 A + \cos^2 A = 1 \qquad\qquad$ (§65)

then $\qquad\qquad \sin^2 A = 1 - \cos^2 A$

and $\qquad\qquad \cos^2 A = 1 - \sin^2 A$

Substituting for $\cos^2 A$ in (4)

$$\cos 2A = 1 - 2 \sin^2 A \qquad\qquad (5)$$

Substituting for $\sin^2 A$

$$\cos 2A = 2 \cos^2 A - 1 \qquad\qquad (6)$$

No. 5 may be written in the form:

$$1 - \cos 2A = 2 \sin^2 A \qquad\qquad (7)$$

and No. 6 as $\qquad 1 + \cos 2A = 2 \cos^2 A \qquad\qquad$ (8)

These alternative forms are very useful.

Again, if (7) be divided by (8)

$$\frac{1 - \cos 2A}{1 + \cos 2A} = \frac{\sin^2 A}{\cos^2 A}$$

or $\qquad\qquad \tan^2 A = \dfrac{1 - \cos 2A}{1 + \cos 2A} \qquad\qquad$ (9)

If $2A$ be replaced by θ, formulae (4), (5) and (6) may be written in the forms

$$\cos \theta = \cos^2 \frac{\theta}{2} - \sin^2 \frac{\theta}{2} \qquad\qquad (10)$$

$$\cos \theta = 1 - 2 \sin^2 \frac{\theta}{2} \qquad\qquad (11)$$

$$\cos \theta = 2 \cos^2 \frac{\theta}{2} - 1 \qquad\qquad (12)$$

84. Similar formulae may be found for tangents.

Since $\qquad \tan (A + B) = \dfrac{\tan A + \tan B}{1 - \tan A \tan B}$

Let $B = A$

Then $\qquad\qquad \tan 2A = \dfrac{2 \tan A}{1 - \tan^2 A} \qquad\qquad$ (13)

or replacing $2A$ by θ

$$\tan \theta = \frac{2 \tan \dfrac{\theta}{2}}{1 - \tan^2 \dfrac{\theta}{2}} \qquad\qquad (14)$$

Formula (11) above may be written in the form:

$$\sin^2 \frac{\theta}{2} = \tfrac{1}{2}(1 - \cos \theta)$$

It is frequently used in Navigation.

$(1 - \cos \theta)$ is called the **versed sine** of θ
and $(1 - \sin \theta)$ is called the **coversed sine** of θ.

$\tfrac{1}{2}(1 - \cos \theta)$ is called the " **haversine** ", *i.e.* half the versed sine.

85. The preceding formulae are so important that they are collected here for future reference.

(1) $\sin (A + B) = \sin A \cos B + \cos A \sin B$
(2) $\sin (A - B) = \sin A \cos B - \cos A \sin B$
(3) $\cos (A + B) = \cos A \cos B - \sin A \sin B$
(4) $\cos (A - B) = \cos A \cos B + \sin A \sin B$
(5) $\tan (A + B) = \dfrac{\tan A + \tan B}{1 - \tan A \tan B}$.
(6) $\tan (A - B) = \dfrac{\tan A - \tan B}{1 + \tan A \tan B}$.
(7) $\sin 2A = 2 \sin A \cos A$
(8) $\cos 2A = \cos^2 A - \sin^2 A$.
$\qquad = 1 - 2 \sin^2 A$
$\qquad = 2 \cos^2 A - 1$
(9) $\tan 2A = \dfrac{2 \tan A}{1 - \tan^2 A}$.

These formulae should be carefully memorised. Variations of (7), (8), (9) in the form θ and $\dfrac{\theta}{2}$ should also be remembered.

Exercise 14

1. If $\sin A = \tfrac{3}{5}$, find $\sin 2A$, $\cos 2A$ and $\tan 2A$.
2. Find $\sin 2\theta$, $\cos 2\theta$, $\tan 2\theta$, when $\sin \theta = 0.25$.
3. Given the values of $\sin 45°$ and $\cos 45°$ deduce the values of $\sin 90°$ and $\cos 90°$ by using the above formulae.
4. If $\cos B = 0.66$, find $\sin 2B$ and $\cos 2B$.
5. Find the values of (1) $2 \sin 36° \cos 36°$.
$\qquad$ (2) $2 \cos^2 36° - 1$.
6. If $\cos 2A = \tfrac{3}{5}$, find $\tan A$.
(Hint.—Use formulae of § 83.)
7. Prove that $\sin \dfrac{\theta}{2} = \pm \sqrt{\dfrac{1 - \cos \theta}{2}}$
$\qquad \cos \dfrac{\theta}{2} = \pm \sqrt{\dfrac{1 + \cos \theta}{2}}$.

8. If $\cos \theta = \frac{1}{2}$, find $\sin \dfrac{\theta}{2}$ and $\cos \dfrac{\theta}{2}$.

(Hint.—Use the results of the previous question.)

9. If $1 - \cos 2\theta = 0.72$, find $\sin \theta$ and check by using the tables.

10. Prove that $\cos^4 \theta - \sin^4 \theta = \cos 2\theta$.

(Hint.—Factorise the left-hand side.)

11. Prove that $\left(\sin \dfrac{\theta}{2} + \cos \dfrac{\theta}{2}\right)^2 - 1 = \sin \theta$.

12. Find the value of $\sqrt{\dfrac{1 - \cos 30°}{1 + \cos 30°}}$.

(Hint.—See formula of § 83.)

86. Product formulae.

The formulae of § 80 give rise to another set of results involving the product of trigonometrical ratios.

We have seen that:

$$\sin (A + B) = \sin A \cos B + \cos A \sin B \qquad (1)$$
$$\sin (A - B) = \sin A \cos B - \cos A \sin B \qquad (2)$$
$$\cos (A + B) = \cos A \cos B - \sin A \sin B \qquad (3)$$
$$\cos (A - B) = \cos A \cos B + \sin A \sin B \qquad (4)$$

Adding (1) and (2)
$$\sin (A + B) + \sin (A - B) = 2 \sin A \cos B$$
Subtracting
$$\sin (A + B) - \sin (A - B) = 2 \cos A \sin B$$
Adding (3) and (4)
$$\cos (A + B) + \cos (A - B) = 2 \cos A \cos B$$
Subtracting
$$\cos (A + B) - \cos (A - B) = - 2 \sin A \sin B$$

These can be written in the forms

$$2 \sin A \cos B = \sin (A + B) + \sin (A - B) \qquad (5)$$
$$2 \cos A \sin B = \sin (A + B) - \sin (A - B) \qquad (6)$$
$$2 \cos A \cos B = \cos (A + B) + \cos (A - B) \qquad (7)$$
$$2 \sin A \sin B = \cos (A - B) - \cos (A + B) \qquad (8)$$

Note.—The order on the right-hand side of (8) must be carefully noted.

87.

Let $\qquad A + B = P$
and $\qquad A - B = Q$
Adding $\qquad 2A = P + Q$
Subtracting $\qquad 2B = P - Q$
$$\therefore \quad A = \dfrac{P + Q}{2}$$
$$B = \dfrac{P - Q}{2}.$$

Substituting in (5), (6), (7) and (8)

$$\sin P + \sin Q = 2 \sin \frac{P+Q}{2} \cos \frac{P-Q}{2} \qquad (9)$$

$$\sin P - \sin Q = 2 \cos \frac{P+Q}{2} \sin \frac{P-Q}{2} \qquad (10)$$

$$\cos P + \cos Q = 2 \cos \frac{P+Q}{2} \cos \frac{P-Q}{2} \qquad (11)$$

$$\cos Q - \cos P = 2 \sin \frac{P+Q}{2} \sin \frac{P-Q}{2} \qquad (12)$$

The formulae (5), (6), (7), (8) enable us *to change the product of two ratios into a sum.*

Formulae (9), (10), (11), (12) enable us *to change the sum of two ratios into a product.*

Again note carefully the order in (12).

88. Worked examples.

Example I. *Express as the sum of two trigonometrical ratios* sin 5θ cos 3θ.

Using $2 \sin A \cos B = \sin (A + B) + \sin (A - B)$
on substitution

$$\sin 5\theta \cos 3\theta = \tfrac{1}{2} \{\sin (5\theta + 3\theta) + \sin (5\theta - 3\theta)\}$$
$$= \tfrac{1}{2} \{\sin 8\theta + \sin 2\theta\}$$

Example 2. *Change into a sum* sin 70° sin 20°.

Using
$$2 \sin A \sin B = \cos (A - B) - \cos (A + B)$$
on substitution

$$\sin 70° \sin 20° = \tfrac{1}{2} \{\cos (70° - 20°) - \cos (70° + 20°)\}$$
$$= \tfrac{1}{2} \{\cos 50° - \cos 90°\}$$
$$= \tfrac{1}{2} \cos 50° \quad \text{since } \cos 90° = 0.$$

Example 3. *Transform into a product* sin 25° + sin 18°.

Using
$$\sin P + \sin Q = 2 \sin \frac{P+Q}{2} \cos \frac{P-Q}{2}$$

$$\sin 25° + \sin 18° = 2 \sin \frac{25° + 18°}{2} \cos \frac{25° - 18°}{2}$$
$$= 2 \sin 21° 30' \cos 3° 30'.$$

Example 4. *Change into a product* cos 3θ − cos 7θ.

Using
$$\cos Q - \cos P = 2 \sin \frac{P+Q}{2} \sin \frac{P-Q}{2}$$

on substitution

$$\cos 3\theta - \cos 7\theta = 2 \sin \frac{3\theta + 7\theta}{2} \sin \frac{7\theta - 3\theta}{2}$$
$$= 2 \sin 5\theta \sin 2\theta.$$

Exercise 15

Express as the *sum* or difference of two ratios:

1. $\sin 3\theta \cos \theta$.
2. $\sin 35° \cos 45°$.
3. $\cos 50° \cos 30°$.
4. $\cos 5\theta \sin 3\theta$.
5. $\cos (C + 2D) \cos (2C + D)$.
6. $\cos 60° \sin 30°$.
7. $2 \sin 3A \sin A$.
8. $\cos (3C + 5D) \sin (3C - 5D)$.

Express as the product of two ratios:

9. $\sin 4A + \sin 2A$.
10. $\sin 5A - \sin A$.
11. $\cos 4\theta - \cos 2\theta$.
12. $\cos A - \cos 5A$.
13. $\cos 47° + \cos 35°$.
14. $\sin 49° - \sin 23°$.
15. $\dfrac{\sin 30° + \sin 60°}{\cos 30° - \cos 60°}$.
16. $\dfrac{\sin \alpha + \sin \beta}{\cos \alpha + \cos \beta}$.

CHAPTER VII

RELATIONS BETWEEN THE SIDES AND ANGLES OF A TRIANGLE

89. In § 61 we considered the relations which exist between the sides and angles of a right-angled triangle. In this Chapter we proceed to deal similarly with any triangle.

In accordance with the usual practice, the angles of a triangle will be denoted by A, B, and C, and the sides *opposite* to these by a, b, and c, respectively.

Note.—In working examples in this and the following chapters, the student will constantly be using logarithms and trigonometrical ratios taken from the tables. It should be

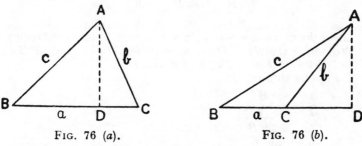

FIG. 76 (*a*).   FIG. 76 (*b*).

remembered that the numbers in these tables are given correct to *four* significant figures only. When they are used in a number of successive operations there will sometimes be an accumulation of small errors which will result in small differences in the answers. In general a three-figure accuracy is all that can be relied upon.

For a general treatment of these errors of approximations the student should consult a good modern arithmetic or a special chapter on them in *National Certificate Mathematics*, Vol. I, published by the English Universities Press.

90. The sine rule.

In every triangle the sides are proportional to the sines of the opposite angles.

There are two cases to be considered:

(1) Acute-angled triangle (Fig. 76(*a*)).
(2) Obtuse-angled triangle (Fig. 76(*b*)).

In each figure draw AD perpendicular to BC, or to BC produced (Fig. 76(b)).

In $\qquad\qquad \triangle ABD, AD = c \sin B$ $\qquad\qquad$ (1)
In $\qquad\qquad \triangle ACD, AD = b \sin C$ $\qquad\qquad$ (2)

In Fig. 76(b), since ACB and ACD are supplementary angles

$$\sin ACD = \sin ACB = \sin C.$$

Equating (1) and (2):

$$c \sin B = b \sin C$$
$$\therefore \quad \frac{b}{c} = \frac{\sin B}{\sin C}$$

Similarly $\qquad\qquad \dfrac{a}{b} = \dfrac{\sin A}{\sin B}$

and $\qquad\qquad \dfrac{a}{c} = \dfrac{\sin A}{\sin C}.$

These results may be combined in the one formula

$$\frac{\sin A}{a} = \frac{\sin B}{b} = \frac{\sin C}{c}$$

These formulæ are suitable for logarithmic calculations.

Worked example. *If in a triangle ABC, A = 52° 15′, B = 70° 26′ and a = 9·8, find b and c.*

Using the sine rule:

$$\frac{b}{a} = \frac{\sin B}{\sin A}$$
$$\therefore \quad b = \frac{a \sin B}{\sin A}$$
$$\therefore \quad \log b = \log a + \log \sin B - \log \sin A$$
$$= \log 9\cdot8 + \log \sin 70° 26′ - \log \sin 52° 15′$$
$$= 0\cdot9912 + \bar{1}\cdot9742 - \bar{1}\cdot8980$$
$$= 1\cdot0674$$
$$= \log 11\cdot68$$
$$\therefore \quad \boldsymbol{b} = 11\cdot7 \text{ (approx.)}$$

Similarly c may be found by using $\dfrac{c}{a} = \dfrac{\sin C}{\sin A}.$

Exercise 16

Solve the following problems connected with a triangle ABC.

1. When $A = 54°$, $B = 67°$, $a = 13\cdot9$ m, find b and c.
2. When $A = 38° 15′$, $B = 29° 38′$, $b = 16\cdot2$ m, find a and c.

3. When $A = 70°$, $C = 58° 16'$, $b = 6$ mm, find a and c.
4. When $A = 88°$, $B = 36°$, $a = 9.5$ m, find b and c.
5. When $B = 75°$, $C = 42°$, $b = 25$ cm, find a and c.

91. The cosine rule.

As in the case of the sine rule, there are two cases to be considered. These are shown in Figs. 77(a) and 77(b).

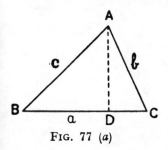

FIG. 77 (a)   FIG. 77 (b).

Let $BD = x$
Then $CD = a - x$ in Fig. 77(a)
and $CD = x - a$ in Fig. 77(b)
In $\triangle ABD$, $AD^2 = AB^2 - BD^2$
                   $= c^2 - x^2$ (1)
In $\triangle ACD$, $AD^2 = AC^2 - CD^2$
                   $= b^2 - (a - x)^2$ in Fig. 77(a)   (2)
or $= b^2 - (x - a)^2$ in Fig. 77(b)
Also $(a - x)^2 = (x - a)^2$
$\therefore$ equating (1) and (2)
                   $b^2 - (a - x)^2 = c^2 - x^2$
$\therefore$ $b^2 - a^2 + 2ax - x^2 = c^2 - x^2$
$\therefore$ $2ax = a^2 + c^2 - b^2$.
But $x = c \cos B$
$\therefore$ $2ac \cos B = a^2 + c^2 - b^2$

$$\therefore \cos B = \frac{a^2 + c^2 - b^2}{2ac}.$$

Similarly

$$\cos A = \frac{b^2 + c^2 - a^2}{2bc}$$

$$\cos C = \frac{a^2 + b^2 - c^2}{2ab}.$$

The formulae may also be written in the forms:
$$c^2 = a^2 + b^2 - 2ab \cos C.$$
$$a^2 = b^2 + c^2 - 2bc \cos A.$$
$$b^2 = a^2 + c^2 - 2ac \cos B.$$

These formulae enable us to find the angles of a triangle when all the sides are known. In the second form it enables us to find the third side when two sides and the enclosed angle are known.

Worked example.

Find the angles of the triangle whose sides are

$$a = 8 \text{ m}, b = 9 \text{ m}, c = 12 \text{ m}.$$

Using
$$\cos C = \frac{a^2 + b^2 - c^2}{2ab}$$

$$= \frac{8^2 + 9^2 - 12^2}{2 \times 8 \times 9}$$

$$= \frac{64 + 81 - 144}{2 \times 8 \times 9}$$

$$= \frac{1}{144}$$

$$= 0 \cdot 0069$$

whence $\qquad C = 89° \ 36'.$

Again, $\qquad \cos A = \dfrac{b^2 + c^2 - a^2}{2bc}$

$$= \frac{9^2 + 12^2 - 8^2}{2 \times 9 \times 12}$$

$$= \frac{81 + 144 - 64}{2 \times 9 \times 12}$$

$$= \frac{161}{216}$$

$$= 0 \cdot 7454$$

whence $\qquad A = 41° \ 48'.$

Similarly, using
$$\cos B = \frac{a^2 + c^2 - b^2}{2ac}$$

we get $\qquad B = 48° \ 36'.$

Check

$$A + B + C$$
$$= 41° \ 48' + 48° \ 36' + 89° \ 36'$$
$$= 180°.$$

Exercise 17

Find the angles of the triangles in which:

1. $a = 2$ km, $b = 3$ km, $c = 4$ km.
2. $a = 54$ mm, $b = 71$ mm, $c = 83$ mm.
3. $a = 24$ m, $b = 19$ m, $c = 26$ m.
4. $a = 2 \cdot 6$ km, $b = 2 \cdot 85$ km, $c = 4 \cdot 7$ km.

5. If $a = 14$ m, $b = 8\cdot5$ m, $c = 9$ m, find the greatest angle of the triangle.

6. When $a = 64$ mm, $b = 57$ mm, and $c = 82$ mm, find the smallest angle of the triangle.

92. The half-angle formulae.

The cosine formula is not suitable for use with logarithms and is tedious when the numbers involved are large: it is the basis, however, of a series of other formulae which are easier to manipulate.

93. To express the sines of half the angles in terms of the sides.

As proved in § 91

$$\cos A = \frac{b^2 + c^2 - a^2}{2bc}$$

but $$\cos A = 1 - 2 \sin^2 \frac{A}{2} \qquad \text{(§ 83)}$$

$$\therefore \quad 1 - 2 \sin^2 \frac{A}{2} = \frac{b^2 + c^2 - a^2}{2bc}$$

$$\therefore \quad 2 \sin^2 \frac{A}{2} = 1 - \frac{b^2 + c^2 - a^2}{2bc}$$

$$= \frac{2bc - (b^2 + c^2 - a^2)}{2bc}$$

$$= \frac{2bc - b^2 - c^2 + a^2}{2bc}$$

$$= \frac{a^2 - (b^2 - 2bc + c^2)}{2bc}$$

$$= \frac{a^2 - (b - c)^2}{2bc}.$$

Factorising the numerator

$$2 \sin^2 \frac{A}{2} = \frac{(a + b - c)(a - b + c)}{2bc} \qquad \text{(A)}$$

The " s " notation. To simplify this further we use the " s " notation, as follows:

Let $2s = a + b + c$, *i.e.* the perimeter of the triangle.

Then $$2s - 2a = a + b + c - 2a$$
$$= b + c - a$$

Again $$2s - 2b = a + b + c - 2b$$
$$= a - b + c$$

Similarly $$2s - 2c = a + b + c.$$

These may be written

$$2s = a + b + c \tag{1}$$
$$2(s - a) = b + c - a \tag{2}$$
$$2(s - b) = a - b + c \tag{3}$$
$$2(s - c) = a + b - c \tag{4}$$

From (A) above

$$2 \sin^2 \frac{A}{2} = \frac{(a + b - c)(a - b + c)}{2bc}.$$

Replacing the factors of the numerator by their equivalents in formulae (3) and (4)

we have $\qquad 2 \sin^2 \dfrac{A}{2} = \dfrac{2(s - c) \times 2(s - b)}{2bc}$

Cancelling the " 2's."

$$\sin^2 \frac{A}{2} = \frac{(s - c)(s - b)}{bc}$$

or $\qquad \sin \dfrac{A}{2} = \sqrt{\dfrac{(s - b)(s - c)}{bc}}$

Similarly, $\qquad \sin \dfrac{B}{2} = \sqrt{\dfrac{(s - a)(s - c)}{ac}}$

$$\sin \frac{C}{2} = \sqrt{\frac{(s - a)(s - b)}{ab}}$$

94. To express the cosines of half the angles of a triangle in terms of the sides.

Since $\qquad \cos A = \dfrac{b^2 + c^2 - a^2}{2bc}$

$$1 + \cos A = 1 + \frac{b^2 + c^2 - a^2}{2bc}$$

but $\qquad 1 + \cos A = 2 \cos^2 \dfrac{A}{2} \qquad$ (Chapter VI, § 83)

$$\therefore \quad 2 \cos^2 \frac{A}{2} = 1 + \frac{b^2 + c^2 - a^2}{2bc}$$

$$= \frac{(b^2 + 2bc + c^2) - a^2}{2bc}$$

$$= \frac{(b + c)^2 - a^2}{2bc}$$

$$= \frac{(b + c - a)(b + c + a)}{2bc}$$

(on factorising the numerator)

but $\qquad b + c - a = 2(s - a)$

and $\qquad a + b + c = 2s$

Substituting
$$2 \cos^2 \frac{A}{2} = \frac{2(s-a) \times 2s}{2bc}$$

and
$$\cos^2 \frac{A}{2} = \frac{s(s-a)}{bc}$$

$$\therefore \quad \cos \frac{A}{2} = \sqrt{\frac{s(s-a)}{bc}}$$

Similarly
$$\cos \frac{B}{2} = \sqrt{\frac{s(s-b)}{ac}}$$

$$\cos \frac{C}{2} = \sqrt{\frac{s(s-c)}{ab}}$$

95. **To express the tangents of half the angles of a triangle in terms of the sides.**

Since
$$\tan \frac{A}{2} = \frac{\sin \frac{A}{2}}{\cos \frac{A}{2}}$$

we can substitute for $\sin \frac{A}{2}$ and $\cos \frac{A}{2}$ the expressions found above.

Then
$$\tan \frac{A}{2} = \frac{\sqrt{\frac{(s-b)(s-c)}{bc}}}{\sqrt{\frac{s(s-a)}{bc}}}$$

Simplifying and cancelling
$$\tan \frac{A}{2} = \sqrt{\frac{(s-b)(s-c)}{s(s-a)}}$$

Similarly
$$\tan \frac{B}{2} = \sqrt{\frac{(s-a)(s-c)}{s(s-b)}}$$

and
$$\tan \frac{C}{2} = \sqrt{\frac{(s-a)(s-b)}{s(s-c)}}$$

96. **To express the sine of an angle of a triangle in terms of the sides.**

Since
$$\sin A = 2 \sin \frac{A}{2} \cos \frac{A}{2}$$

substituting for $\sin \frac{A}{2}$ and $\cos \frac{A}{2}$ the values found above

$$\sin A = 2\sqrt{\frac{(s-b)(s-c)}{bc}} \times \sqrt{\frac{s(s-a)}{bc}}$$

$$\therefore \quad \sin A = \frac{2}{bc} \sqrt{s(s-a)(s-b)(s-c)}, \text{ on simplifying.}$$

Similarly

$$\sin B = \frac{2}{ac} \sqrt{s(s-a)(s-b)(s-c)}$$

and

$$\sin C = \frac{2}{ab} \sqrt{s(s-a)(s-b)(s-c)}$$

97. Worked example.

The working involved in the use of all these formulae is very similar. We will give one example only: others will be found in the next chapter.

The sides of a triangle are $a = 264, b = 435, c = 473$. Find the greatest angle.

The greatest angle is opposite to the greatest side and is therefore C.

In questions of this type it is very important to employ a clear and methodical arrangement of the working. Unless this is done loss of time and inaccurate results will follow.

Checks should be employed at suitable stages.

The following arrangement is suggested.

Begin by calculating values of the " s " factors and setting out their logarithms.

			logs.
	$a =$	264	
	$b =$	435	
	$c =$	473	
$\therefore$	$2s =$	1172	
	$s =$	586	2·7679
	$s - a =$	322	2·5079
	$s - b =$	151	2·1790
	$s - c =$	113	2·0531
Check	$2s =$	1172	

and

Note.—$s + (s - a) + (s - b) + (s - c) =$
$$4s - (a + b + c) = 2s.$$

Any of the half angle formulæ may be used, but the tangent formulae involves only the " s " factors, all the logs of which are set out above.

Using

$$\tan \frac{C}{2} = \sqrt{\frac{(s-a)(s-b)}{s(s-c)}}$$

$$\tan \frac{C}{2} = \sqrt{\frac{322 \times 151}{586 \times 113}}$$

$$\therefore \ \log \tan \frac{C}{2} = \tfrac{1}{2}(\log 322 + \log 151 - \log 586 - \log 113)$$

$$= \overline{1}\cdot9329 \qquad \text{(see working)}$$

$$\therefore \ \frac{C}{2} = 40° \ 36'$$

and $\quad C = 81° \ 12'$

No.	Log.
322	2·5079
151	2·1790
	4·6869
586	2·7679
112	2·0531
	4·8210
÷ 2	$\overline{1}$·8659
	$\overline{1}$·9329

Exercise 18

1. Using the formula for $\tan \frac{A}{2}$, find the largest angle in the triangle whose sides are 113 mm, 141 mm, 214 mm.

2. Using the formula for $\sin \frac{A}{2}$, find the smallest angle in the triangle whose sides are 483 mm, 316 mm, and 624 mm.

3. Using the formula for $\cos \frac{B}{2}$ find B when $a = 115$ m, $b = 221$ m, $c = 286$ m.

4. Using the half-angle formulae find the angles of the triangle when $a = 160$, $b = 220$, $c = 340$.

5. Using the half-angle formulae find the angles of the triangle whose sides are 73·5, 65·5 and 75.

6. Using the formula for the sine in § 96 find the smallest angle of the triangle whose sides are 172 km, 208 km, and 274 km.

98. To prove that in any triangle

$$\tan \frac{B-C}{2} = \frac{b-c}{b+c} \cot \frac{A}{2}$$

From § 90 $\qquad \dfrac{\sin B}{b} = \dfrac{\sin C}{c}$

Let each of these ratios equal k.

Then $\hspace{2cm} \sin B = bk$ $\hspace{2cm}$ **(1)**

and $\hspace{2cm} \sin C = ck$ $\hspace{2cm}$ **(2)**

Adding (1) and (2)
$$\sin B + \sin C = k(b + c) \hspace{1cm} \textbf{(3)}$$

Subtracting (2) from (1)
$$\sin B - \sin C = k(b - c) \hspace{1cm} \textbf{(4)}$$

Dividing (4) by (3)
$$\frac{\sin B - \sin C}{\sin B + \sin C} = \frac{b - c}{b + c}$$

or $\hspace{1cm} \dfrac{b - c}{b + c} = \dfrac{\sin B - \sin C}{\sin B + \sin C}$

Applying to the numerator and denominator of the right-hand side the formulae, 9 and 10 of § 87.

We get $\hspace{0.5cm} \dfrac{b - c}{b + c} = \dfrac{2 \cos \frac{B + C}{2} \cdot \sin \frac{B - C}{2}}{2 \sin \frac{B + C}{2} \cdot \cos \frac{B - C}{2}}$

$$= \frac{\sin \frac{B - C}{2}}{\cos \frac{B - C}{2}} \div \frac{\sin \frac{B + C}{2}}{\cos \frac{B + C}{2}}$$

$$= \frac{\tan \frac{B - C}{2}}{\tan \frac{B + C}{2}}$$

Since $\hspace{1cm} (B + C) = 180° - A$

$\therefore \hspace{1cm} \dfrac{B + C}{2} = 90° - \dfrac{A}{2}$

$\therefore \hspace{1cm} \dfrac{b - c}{b + c} = \dfrac{\tan \frac{B - C}{2}}{\tan \left(90° - \frac{A}{2}\right)}$

$$= \frac{\tan \frac{B - C}{2}}{\cot \frac{A}{2}} \hspace{1cm} \text{(see § 53)}$$

$\therefore \hspace{0.5cm} \dfrac{\tan \frac{(B - C)}{2}}{\cot \frac{A}{2}} = \dfrac{b - c}{b + c}$

or $\hspace{1cm} \tan \dfrac{B - C}{2} = \dfrac{b - c}{b + c} \cot \dfrac{A}{2}$

Similarly

$$\tan \frac{A - C}{2} = \frac{a - c}{a + c} \cot \frac{B}{2}$$

$$\tan \frac{A - B}{2} = \frac{a - b}{a + b} \cot \frac{C}{2}$$

This formula is well adapted for use with logarithms, and although at first sight it may look a complicated one it is not difficult to manipulate.

On the right-hand side we have quantities which are known when we are given *two sides of a triangle and the contained angle*.

Consequently we can find $\frac{B - C}{2}$ and so $B - C$.

Since A is known we can find $B + C$ for $B + C = 180 - A$

Let $B + C = \alpha$

,, $B - C = \beta$ (note α and β are now known)

Adding $2B = \alpha + \beta$

Subtracting $2C = \alpha - \beta$

$$\therefore \quad B = \frac{\alpha + \beta}{2} \text{ and } C = \frac{\alpha - \beta}{2}$$

Hence we know all the angles of the triangle.

Worked example.

In a triangle $A = 75° 12'$, $b = 43$, $c = 35$. Find B and C.

Using $\tan \dfrac{B - C}{2} = \dfrac{b - c}{b + c} \cot \dfrac{A}{2}$

and substituting

$$\tan \frac{B - C}{2} = \frac{43 - 35}{43 + 35} \cot 37° 36'$$

$$= \frac{8}{78} \cot 37° 36'$$

$$\therefore \quad \log \tan \frac{B - C}{2} = \log 8 + \log \cot 37° 36' - \log 78$$

$$= \bar{1} \cdot 1245$$

	No.	Log.
	8	0·9031
cot 37° 36′		0·1135
		1·0166
	78	1·8921
log tan 7° 35′		$\bar{1}$·1245

whence $\dfrac{B - C}{2} = 7° 35'$

and $B - C = 15° 10'$

Also $B + C = 180° - 75° 12'$

               $= 104° 48'$

(1) Adding $2B = 119° 58'$

and $B = 59° 59'$

$$\therefore \quad \mathbf{B} = 60° \text{ approx.}$$

(2) Subtracting $2C = 89° 38'$
and $\qquad C = 44° 49'$.

99. To prove that in any triangle

$$a = b \cos C + c \cos B$$

As in § 90 there are two cases.
In Fig. 78(a) $\quad BC = BD + DC$
But $\qquad BD = c \cos B$
and $\qquad DC = b \cos C$
$\quad \therefore \quad a = BD + DC$
$\qquad = c \cos B + b \cos C$

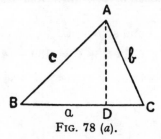

FIG. 78 (a).

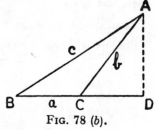

FIG. 78 (b).

In Fig. 78 (b) $\quad BC = BD - DC$
$\quad \therefore \quad a = c \cos B - b \cos ACD$
$\qquad = c \cos B - b \cos (180° - C)$
$\qquad = c \cos B + b \cos C$

since
$\qquad \cos (180° - B) = - \cos B \qquad \text{(see § 70)}$
$\therefore$ in each case

$$a = b \cos C + c \cos B$$

Similarly $\qquad b = a \cos C + c \cos A$
and $\qquad c = a \cos B + b \cos A$

Referring to § 63 we see that BD is the projection of AB on BC, and DC is the projection of AC on BC; in the second case BC is produced and the projection must be regarded as negative. Hence we may state the Theorem thus:

Any side of a triangle is equal to the projection on it of the other two sides.

Exercise 19.

Use the formula proved in § 98 to find the remaining angles of the following triangles:
1. $a = 171$, $c = 288$, $B = 108°$.
2. $a = 786$, $b = 854$, $C = 37° 25'$.
3. $c = 175$, $b = 602$, $A = 63° 40'$.
4. $a = 185$, $b = 111$, $C = 60°$.
5. $a = 431$, $b = 387$, $C = 29° 14'$.
6. $a = 759$, $c = 567$, $B = 72° 14'$.

CHAPTER VIII

THE SOLUTION OF TRIANGLES

100. THE formulae which have been proved in the previous chapter are those which are used for the purpose of *solving a triangle*. By this is meant that, given certain of the sides and angles of a triangle, we proceed to find the others. The parts given must be such as to make it possible to determine the triangle uniquely. If, for example, all the angles are given, there is no *one* triangle which has these angles, but an infinite number of such triangles, with different lengths of corresponding sides. Such triangles are *similar*, but not congruent (see § 15).

The conditions under which the solution of a triangle is possible must be the same as those which determine when triangles are *congruent*. The student, before proceeding further, should revise these conditions (see Chapter I, § 13).

It should be understood, of course, that we are not dealing now with right-angled triangles, which have already been considered (see Chapter III, § 62).

101. From the Theorems enumerated in § 13, it is clear that a triangle can be " *solved* " when the following parts are given:

Case I.   Three sides.
Case II.   Two sides and an included angle.
Case III. Two angles and a side.
Case IV.   Two sides and an angle opposite to one of them.

This last case, however, is the *Ambiguous Case* (see § 13) and under certain conditions, which will be dealt with later, *there may be two solutions*.

In the previous chapter, after proving the various formulae, examples were considered which were, in effect, concerned with the solution of a triangle, but we must now proceed to a systematic consideration of the whole problem.

102. Case I.   To solve a triangle when three sides are known.

The problem is that of finding at least two of the angles,

because since the sum of the angles of a triangle is 180°, when two are known the third can be found by subtraction. It is better, however, to calculate all three angles separately and check the result by seeing if their sum is 180°.

Formulae employed.

(1) **The cosine rule.** The formula

$$\cos A = \frac{b^2 + c^2 - a^2}{2bc}$$

will give A, and B and C can be similarly determined. As previously stated, however, this should only be used if the numbers are small, since it is not suitable for logarithmic calculations.

(2) **The half angle formulae.** The best of these, as previously pointed out, is the $\tan \frac{A}{2}$ formula, viz.

$$\tan \frac{A}{2} = \sqrt{\frac{(s-b)(s-c)}{s(s-a)}}$$

However, the formulae for $\sin \frac{A}{2}$ and $\cos \frac{A}{2}$ may be used.

(3) **The sine formula**

$$\sin A = \frac{2}{bc} \sqrt{s(s-a)(s-b)(s-c)}$$

This is longer than the half-angle formulae, though suitable for logarithmic calculations.

Worked example.

Solve the triangle in which a = 269·8, b = 235·9, c = 264·7.

Data and logs.

		Logs.
$a =$	269·8	
$b =$	235·9	
$c =$	264·7	
$2s =$	770·4	
$\therefore \quad s =$	385·2	2·5857
$s - a =$	115·4	2·0622
$s - b =$	149·3	2·1741
$s - c =$	120·5	2·0809

Check $2s = 770·4$

To find A

Formula to be used $\tan \dfrac{A}{2} = \sqrt{\dfrac{(s-b)(s-c)}{s(s-a)}}$

Taking logs

$\log \tan \dfrac{A}{2} = \dfrac{1}{2}[\{\log(s-b) + \log(s-c)\} - \{\log s + \log(s-a)\}]$

$\qquad = \bar{1}\cdot 8035$ (from working)
$\qquad = \log \tan 32° \ 28'$

$\therefore \quad \dfrac{A}{2} = 32° \ 28'$

and $\quad A = 64° \ 56'$.

	Logs.
$s-b$	$2\cdot1741$
$s-c$	$2\cdot0809$
	$4\cdot2550$
s	$2\cdot5857$
$s-a$	$2\cdot0622$
	$4\cdot6479$
$\div\ 2$	$\bar{1}\cdot6071$
	$\bar{1}\cdot8035$

To find B.

Formula used

$$\tan \dfrac{B}{2} = \sqrt{\dfrac{(s-a)(s-c)}{s(s-b)}}$$

Taking logs

$\log \tan \dfrac{B}{2} = \dfrac{1}{2}[\{\log(s-a) + \log(s-c)\} - \{\log s + \log(s-b)\}]$

$\qquad = \bar{1}\cdot6916$
$\qquad = \log \tan 26° \ 11'$

$\therefore \quad \dfrac{B}{2} = 26° \ 11'$

and $\quad B = 52° \ 22'$.

	Logs.
$s-a$	$2\cdot0622$
$s-c$	$2\cdot0809$
	$4\cdot1431$
s	$2\cdot5857$
$s-b$	$2\cdot1741$
	$4\cdot7598$
$\div\ 2$	$\bar{1}\cdot3833$
	$\bar{1}\cdot6916$

To find C.

Formula used

$$\tan \frac{C}{2} = \sqrt{\frac{(s-a)(s-b)}{s(s-c)}}$$

Taking logs

$$\log \tan \frac{C}{2} = \frac{1}{2}[\{\log(s-a) + \log(s-b)\} - \{\log s + \log(s-c)\}]$$

$$= \bar{1}\cdot7848$$
$$= \log \tan 31° 21'$$

$$\therefore \quad \frac{C}{2} = 31° 21'$$
$$C = 62° 42'.$$

Check

$$A = 64° 56'$$
$$B = 52° 22'$$
$$C = 62° 42'$$

$$\boldsymbol{A + B + C = 180° 00'}$$

	Logs.
$s - a$	2·0622
$s - b$	2·1741
	4·2363
s	2·5857
$s - c$	2·0809
	4·6666
$\div 2$	1·5697
	$\bar{1}$·7848

Exercise 20.

Solve the following triangles:

1. $a = 252$, $b = 342$, $c = 486$.
2. $a = 10$, $b = 11$, $c = 12$.
3. $a = 206\cdot5$, $b = 177$, $c = 295$.
4. $a = 402\cdot5$, $b = 773\cdot5$, $c = 1001$.
5. $a = 95\cdot2$, $b = 162\cdot4$, $c = 117\cdot6$.

103. Case II.   Given two sides and the contained angle.

(1) The *cosine rule* may be used. If, for example, the given sides are b and c and the angle A, then

$$a^2 = b^2 + c^2 - 2bc \cos A$$

will give a.

Hence, since all sides are now known we can proceed as in Case I. The drawbacks to the use of this formula were given in the previous case.

(2) Use the formula

$$\tan \frac{\boldsymbol{B - C}}{2} = \frac{\boldsymbol{b - c}}{\boldsymbol{b + c}} \cot \frac{\boldsymbol{A}}{2}$$

which is suitable for use with logarithms.

Solve the triangle when

$$b = 294, \ c = 406, \ A = 35° 24'$$

Data and logs :

$$
\begin{array}{r|l}
b = 294 & 2\cdot4683 \\
c = 406 & 2\cdot6085 \\
c + b = 700 & 2\cdot8451 \\
{}^*c - b = 112 & 2\cdot0492 \\
A = 35^\circ\,24' & \\
\dfrac{A}{2} = 17^\circ\,42' & 0\cdot4960 \quad \left(\log \cot \dfrac{A}{2}\right) \\
C + B = 144^\circ\,36' &
\end{array}
$$

* This form is used since $c > b$, and therefore $C > B$.

Formula used :

$$\tan \frac{C - B}{2} = \frac{c - b}{c + b} \cot \frac{A}{2}$$

$$\therefore \quad \log \tan \frac{C - B}{2} = \log(c - b) + \log \cot \frac{A}{2} - \log(c + b)$$

$$= \bar{1}\cdot7001$$

$$= \log \tan 26^\circ\,38'$$

		Logs.
$\therefore \quad \dfrac{C - B}{2} = 26^\circ\,38'$	112	2·0492
	cot $17^\circ\,42'$	0·4960
$C - B = 53^\circ\,16'$		2·5452
Also $C + B = 144^\circ\,36'$	700	2·8451
$2C = 197^\circ\,52'$		
$C = 98^\circ\,56'$	tan $26^\circ\,38'$	$\bar{1}$·7001
Also $2B = 91^\circ\,20'$		
$B = 45^\circ\,40'.$		

To find **a**.

Formula used :

$$\frac{a}{\sin A} = \frac{b}{\sin B}$$

$$\therefore \quad \log a = \log b + \log \sin A - \log \sin B$$

$$= 2\cdot3767$$

$$= \log 238\cdot1$$

$$\therefore \quad a = 238 \text{ approx.}$$

The solution is :

		Logs.
	294	2·4683
	sin $35^\circ\,24'$	$\bar{1}$·7629
$B = 45^\circ\,40'$		2·2312
$C = 98^\circ\,56'$	sin $45^\circ\,40'$	$\bar{1}$·8545
$a = 238.$		
	238	2·3767

Exercise 21

Solve the following triangles :
1. $b = 189$, $c = 117\cdot7$, $A = 60^\circ\,36'$.
2. $a = 94$, $b = 159\cdot4$, $C = 80^\circ\,58'$.
3. $a = 39\cdot6$, $c = 71\cdot1$, $B = 65^\circ\,10'$.
4. $a = 266$, $b = 175$, $C = 78^\circ$.
5. $a = 230\cdot1$, $c = 269\cdot5$, $B = 30^\circ\,28'$.

104. Case III. Given two angles and a side.

If two angles are known the third is also known, since the sum of all three angles is 180°. This case may therefore be stated as

Given the angles and one side.

The best formula to use is the *Sine rule*.

Note.—It has previously been stated that if greater accuracy is required than can be obtained by the use of four-figure tables, a book giving seven-figure tables is necessary. In order that the student may have some idea of these tables and their use, they will be employed in the following worked example. Many students will certainly need these more exact tables when they apply their trigonometry to practical problems; they are therefore advised to obtain a copy of *Chambers'* " *Tables* ". The use of them differs in some respects from those employed in four-figure tables, but a full explanation is given in an introduction to the book itself.

Worked example.

Solve the triangle in which $B = 71° 19' 5''$, $C = 67° 27' 33''$ *and* $b = 79·063$.

Note.—It will be observed that the angles are given to the " nearest second " and the length of the side to 5 significant figures.

Required to find, A, a and c.

Now $A = 180° - (71° 19' 5'' + 67° 27' 33'')$
$= 41° 13' 22''$.

To find c.

Formula used $\qquad \dfrac{c}{b} = \dfrac{\sin C}{\sin B}$

whence $\qquad\qquad c = \dfrac{b \sin c}{\sin B}$

$\therefore \quad \log c = \log b + \log \sin c - \log \sin B$

$\qquad\qquad = 1·8869718$
$\qquad\qquad = \log 77·085$
$\therefore \quad C = 77·085.$

	Logs.
79·063	1·8979775
sin 67° 27' 33''	$\bar{1}$·9654810
	1·8634645
sin 71° 19' 5''	$\bar{1}$·9764927
77·085	1·8869718

To find a.

Using $\quad \dfrac{a}{b} = \dfrac{\sin A}{\sin B}$

$\log a = \log b + \log \sin A - \log \sin B$

$\qquad = 1 \cdot 7403627$

$\qquad = \log 55$

$\therefore \quad \boldsymbol{a} = 55.$

	Logs.
79·063	1·8979775
sin 41° 13′ 22″	$\overline{1}$·8188779
	1·7168554
sin 71° 19′ 5″	$\overline{1}$·9764927
55	1·7403627

$\therefore$ The solution is

$$A = 41° 13′ 22″$$
$$\boldsymbol{a} = 55$$
$$\boldsymbol{c} = 77 \cdot 085.$$

Exercise 22

Solve the triangles:

1. $a = 141 \cdot 4$, $A = 74° 18′$, $C = 24° 14′$.
2. $b = 208 \cdot 5$, $A = 95° 41′$, $B = 41° 38′$.
3. $A = 29° 56′$, $C = 108°$, $a = 112 \cdot 8$.
4. $B = 32° 41′$, $C = 49° 38′$, $c = 117 \cdot 6$.
5. $b = 11 \cdot 74$, $A = 27° 45′$, $B = 41° 22′$.

105. Case IV.   Given two sides and an angle opposite to one of them.

This is the *ambiguous case* and the student is advised to revise Chapter I, § 13, before proceeding further.

As we have seen if two sides and an angle *opposite* to one of them be given, then the triangle is not always *uniquely* determined as in the previous cases, but there may be two solutions.

We will now consider from a trigonometrical point of view how this ambiguity may arise.

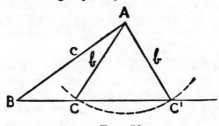

FIG. 79.

In the $\triangle ABC$ (Fig. 79), let c, b, B be known.

As previously shown in § 13 the side b may be drawn in two positions AC and AC'.

Both the triangles ABC and ABC' satisfy the given conditions. Consequently there are:

(1) Two values for a, viz. BC and BC'.
(2) Two values for $\angle C$, viz. ACB or $AC'B$.
(3) Two values for $\angle A$, viz. BAC or BAC'.

Now the $\triangle ACC'$ is isosceles, since $AC = AC'$

$$\therefore \quad \angle ACC' = AC'C.$$

But ACC' is the supplement of ACB.

$\therefore$ also $AC'C$ is the supplement of ACB.

$\therefore$ the *two possible values* of $\angle C$, viz. ACB *and* $AC'B$ *are supplementary.*

Solution.

Since c, b, B are known, C can be found by the sine rule.

i.e. we use
$$\frac{\sin C}{c} = \frac{\sin B}{b}$$

whence
$$\sin C = \frac{c \sin B}{b}$$

Let us suppose that $c = 8\cdot7$, $b = 7\cdot6$, $B = 25$.

Then $\sin C = \dfrac{8\cdot7 \sin 25°}{7\cdot6}$

$\therefore \log \sin C = \log 8\cdot7 + \log \sin 25°$
$\qquad\qquad\qquad - \log 7\cdot6$

$\therefore \log \sin C = \bar{1}\cdot6846.$

	Logs.
$8\cdot7$	$0\cdot9395$
$\sin 25°$	$\bar{1}\cdot6259$
	$0\cdot5654$
$\log 7\cdot6$	$0\cdot8808$
	$\bar{1}\cdot6846$

We have seen in § 73 that when the value of a sine is given, there are two angles less than 180° which have that sine, and the angles are supplementary. Now from the tables the acute angle whose log sine is $\bar{1}\cdot6846$ is 28° 56'.

$\therefore$ $\bar{1}\cdot6846$ is also the log sine of 180° − 28° 36', *i.e.* 151° 4. Consequently there are two values for C, viz.

28° 56' and 151° 4'.

Let us examine the question further by considering the consequences of variations relative to c in the value of b, the side opposite to the given angle B.

As before draw BA making the given angle B meet BX, of indefinite length. Then with centre A and radius $= b$ draw an arc of a circle.

(1) If this arc *touches* BX in C, we have the minimum length of b to make a triangle at all (Fig. 80(*a*)). The triangle is then right-angled, there is *no ambiguity* and

$$b = c \sin B.$$

(2) If b is $> c \sin B$ but $< c$ then BX is cut in two points C and C' (Fig. 80(*b*)).

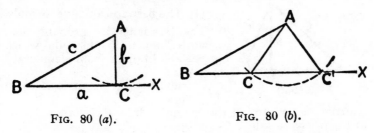

FIG. 80 (*a*).   FIG. 80 (*b*).

There are two △s ABC, ABC' and the *case is ambiguous*.

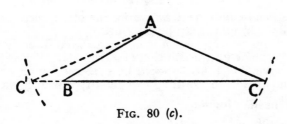

FIG. 80 (*c*).

(3) If $b > c$, BX is cut at two points C and C' (Fig. 80(*c*)), but one of these C' lies on BX produced in the other direction and in the △ so formed, there is no angle B, but only its supplement. There is *one solution* and no *ambiguity*.

∴   *There are two solutions only when b, the side opposite to the given angle B, is less than c, the side adjacent, and greater than c sin B.*

Ambiguity can therefore be ascertained by inspection.

Exercise 23.

In the following cases ascertain if there is more than one solution.   Then solve the triangles:

1. $b = 30{\cdot}4$, $c = 34{\cdot}8$, $B = 25°$.
2. $b = 70{\cdot}25$, $c = 85{\cdot}3$, $B = 40°$.
3. $a = 96$, $c = 100$, $C = 66°$.
4. $a = 91$, $c = 78$, $C = 29° 27'$.

106. Area of a triangle.

From many practical points of view, *e.g.* surveying, the calculation of the area of a triangle is an essential part of solving the triangle. This can be done more readily when the sides and angles are known. This will be apparent in the following formulae.

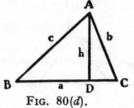

FIG. 80(*d*).

(1) The base and altitude formula.

The student is probably acquainted with this formula which is easily obtained from elementary geometry.

Considering the triangle *ABC* in Fig. 80(*d*).

From *A*, a vertex of the triangle, draw *AD* perpendicular to the opposite side.

Let $AD = h$ and let $\triangle =$ the area of the triangle.
Then
$$\triangle = \tfrac{1}{2}BC \times AD$$
$$= \tfrac{1}{2}ah.$$

If perpendiculars be drawn from the other vertices *B* and *C*, similar formulae may be obtained.

It will be noticed that *h* is not calculated directly in any of the formulae for the solution of a triangle. It is generally more convenient that it should be expressed in terms of the sides and angle. Accordingly we modify this formula in (2).

(2) The sine formula.

Referring to Fig. 80(*d*):
$$\frac{AD}{AC} = \sin C$$
$$\therefore \quad h = b \sin C$$

Substituting for *h* in formula above,
$$\triangle = \tfrac{1}{2}ab \sin C$$

Similarly using other sides as bases
$$\triangle = \tfrac{1}{2}bc \sin A$$
$$= \tfrac{1}{2}ac \sin B.$$

This is a useful formula and adapted to logarithmic calculation. It may be expressed as follows:

The area of a triangle is equal to half the product of two sides and the sine of the angle contained by them.

(3) Area in terms of the sides.

We have seen in § 96, Chapter VII, that
$$\sin A = \frac{2}{bc} \sqrt{s(s-a)(s-b)(s-c)}$$

Substituting this for $\sin A$ in the formula

$$\triangle = \tfrac{1}{2}bc \sin A$$

$$\triangle = \tfrac{1}{2}bc \times \frac{2}{bc} \sqrt{s(s-a)(s-b)(s-c)}$$

$$\therefore \quad \triangle = \sqrt{s(s-a)(s-b)(s-c)}$$

In using this formula with logs the student should revise the hints given in the worked example in § 97, Chapter VII.

Worked examples.

(1) *Find the area of the triangle solved in* § 103, *viz.* $b = 294$, $c = 406$, $A = 35° 24'$.

Using the formula:

$$\triangle = \tfrac{1}{2}bc \sin A$$
$$\triangle = \tfrac{1}{2} \times 294 \times 406 \times \sin 35° 24'$$
$$\log \triangle = \log (0\cdot5) + \log 294 + \log 406 + \log \sin 35° 24'$$
$$= 4\cdot5387$$
$$\therefore \quad \triangle = 34570 \text{ sq. units.}$$

	Logs.
$0\cdot5$	$\bar{1}\cdot6990$
294	$2\cdot4683$
406	$2\cdot6085$
$\sin 35° 24'$	$\bar{1}\cdot7629$
34570	$4\cdot5387$

(2) *Find the area of the triangle solved in* § 102, *viz.* $a = 269\cdot8$, $b = 235\cdot9$, $c = 264\cdot7$.

Using the formula and taking values of s, $s-a$, etc., as in § 102:

$$\triangle = \sqrt{s(s-a)(s-b)(s-c)}$$
$$\therefore \quad \log \triangle = \tfrac{1}{2}\{\log s + \log(s-a) + \log(s-b) + \log(s-c)$$
$$= 4\cdot4515$$
$$= \log 28280$$
$$\therefore \quad \triangle = 28280 \text{ sq. units.}$$

	Log.
$s = 385\cdot2$	$2\cdot5857$
$s - a = 115\cdot4$	$2\cdot0622$
$s - b = 149\cdot3$	$2\cdot1741$
$s - c = 120\cdot5$	$2\cdot0809$
$\div 2$	$8\cdot9029$
28280	$4\cdot4515$

Exercise 24

1. Find the area of the triangle when $a = 6\cdot2$ m, $b = 7\cdot8$ m, $C = 52°$.

2. Find the area of the triangle ABC when $AB = 14$ km, $BC = 11$ km and $\angle ABC = 70°$.

3. If the area of a triangle is 100 m² and two of its sides are 21 m and 15 m, find the angle between these sides.

4. Find the area of the triangle when $a = 98\cdot2$ cm, $c = 73\cdot5$ cm and $B = 135°\ 20'$.

5. Find the area of the triangle whose sides are 28·7 cm, 35·4 cm and 51·8 cm.

6. The sides of a triangle are 10 mm, 13 mm and 17 mm. Find its area.

7. Find the area of the triangle whose sides are 23·22, 31·18 and 40·04 Mm.

8. Find the area of the triangle whose sides are 325 m, 256 m and 189 m.

9. A triangle whose sides are 135 μm, 324 μm and 351 μm is made of material whose density is 2·3 kg μm⁻². Find the mass of the triangle in Mg.

10. Find the area of a quadrilateral $ABCD$, in which $AB = 14\cdot7$ cm, $BC = 9\cdot8$ cm, $CD = 21\cdot7$ cm, $AD = 18\cdot9$ cm and $\angle ABC = 137°$.

11. ABC is a triangle with sides $BC = 36$ cm, $CA = 25$ cm, $AB = 29$ cm. A point O lies inside the triangle and is distant 5 cm from BC and 10 cm from CA. Find its distance from AB.

Exercise 25

Miscellaneous Examples

1. The least side of a triangle is 3·6 km long. Two of the angles are 37° 15' and 48° 24'. Find the greatest side.

2. The sides of a triangle are 123 m, 79 m and 97 m. Find its angles as accurately as you can.

3. Given $b = 532\cdot4$, $c = 647\cdot1$, $A = 75°\ 14'$, find B, C and a.

4. In a triangle ABC find the angle ACB when $AB = 92$ mm, $BC = 50$ mm and $CA = 110$ mm.

5. The length of the side BC of a triangle ABC is 14·5 m $\angle ABC = 71°$, $\angle BAC = 57°$. Calculate the lengths of the sides AC and AB.

6. In a quadrilateral $ABCD$, $AB = 3$ m, $BC = 4$ m, $CD = 7\cdot4$ m, $DA = 4\cdot4$ m and the $\angle ABC$ is 90°. Determine the angle ADC.

7. When $a = 25$, $b = 30$, $A = 50°$ determine how many such triangles exist and complete their solution.

8. The length of the shortest side of a triangle is 162 m. If two angles are 37° 15' and 48° 24' find the greatest side.

9. In a quadrilateral $ABCD$, $AB = 4\cdot3$ m, $BC = 3\cdot4$ m, $CD = 3\cdot8$ m, $\angle ABC = 95°$, $\angle BCD = 115°$. Find the lengths of the diagonals.

10. From a point O on a straight line OX, OP and OQ of lengths 5 mm and 7 mm are drawn on the same side of OX so that $\angle XOP = 32°$ and $\angle XOQ = 55°$. Find the length of PQ.

11. Two hooks P and Q on a horizontal beam are 30 cm apart. From P and Q strings PR and QR, 18 cm and 16 cm long respectively, support a weight at R. Find the distance of R from the beam and the angles which PR and QR make with the beam.

12. Construct a triangle ABC whose base is 5 cm long, the angle $BAC = 55°$ and the angle $ABC = 48°$. Calculate the lengths of the sides AC and BC and the area of the triangle.

13. Two ships leave port at the same time. The first steams S.E. at 18 km h^{-1}, and the second 25° W. of S. at 15 km h^{-1}. Calculate the time that will have elapsed when they are 86 km apart.

14. AB is a base line of length 3 km, and C, D are points such that $\angle BAC = 32° 15'$, $\angle ABC = 119° 5'$, $\angle DBC = 60° 10'$, $\angle BCD = 78° 45'$, A and D being on the same side of BC. Prove that the length of CD is 4405 m approximately.

15. $ABCD$ is a quadrilateral. If $AB = 0.38$ m, $BC = 0.69$ m, $AD = 0.42$ m, $\angle ABC = 109°$, $\angle BAD = 123°$, find the area of the quadrilateral.

16. A weight was hung from a horizontal beam by two chains 8 m and 9 m long respectively, the ends of the chains being fastened to the same point of the weight, their other ends being fastened to the beam at points 10 m apart. Determine the angles which the chains make with the beam.

CHAPTER IX

PRACTICAL PROBLEMS INVOLVING THE SOLUTION OF TRIANGLES

107. It is not possible within the limits of this book to deal with the many practical applications of Trigonometry. For adequate treatment of these the student must consult the technical treatises specially written for those professions in which the subject is necessary. All that is attempted in this chapter is the consideration of a few types of problems which embody those principles which are common to most of the technical applications. Exercises are provided which will provide a training in the use of the rules and formulae which have been studied in previous chapters. In other words, the student must learn to use his tools efficiently and accurately.

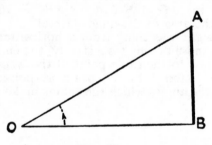

FIG. 81.

108. Determination of the height of a distant object.

This problem has occupied the attention of mankind throughout the ages and is not less important in these days of aeroplanes and balloons. Three simple forms of the problem may be considered here.

(a) **When the point vertically beneath the top of the object is accessible.**

In Fig. 81 *AB* represents a lofty object whose height is required, and *B* is the foot of it, on the same horizontal level as *O*. This being accessible a horizontal distance represented by *OB* can be measured. By the aid of a theodolite the angle of elevation of AB, viz. $\angle AOB$, can be found.

Then $\qquad AB = OB \tan AOB.$

The case of the pyramid considered in Chapter III, § 40, is an example of this. It was assumed that distance from the point vertically below the top of the pyramid could be found.

(b) **When the point on the ground vertically beneath the top of the object is not accessible.**

In Fig. 82 AB represents the height to be determined and B is not accessible. To determine AB we can proceed as follows :

From a suitable point Q, $\angle AQB$ *is measured* by means of a theodolite.

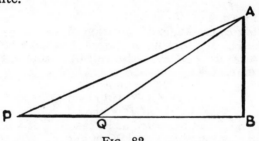

Fig. 82.

Then a distance PQ *is measured* so that P and Q are on the same horizontal plane as B and the $\triangle APQ$ and AB are in the same vertical plane.

Then $\angle APQ$ *is measured*.

∴ in $\triangle APQ$.

PQ is known.

$\angle APQ$ is known.

$\angle AQP$ is known, being the supplement of $\angle AQB$.

The $\triangle APQ$ can therefore be solved as in Case III, § 104.

When AP is known.

Then $\qquad AB = AP \sin APB$

As a check $\qquad AB = AQ \sin AQB$

(c) **By measuring a horizontal distance in any direction.**

It is not always easy to obtain a distance PQ as in the previous example, so that $\triangle APQ$ and AB are in the same vertical plane.

The following method can then be employed.

In Fig. 83 let AB represent the height to be measured.

Taking a point P, measure a horizontal distance PQ in *any* suitable direction.

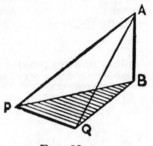

FIG. 83.

At P measure

(1) $\angle APB$, the angle of elevation of A.

(2) $\angle APQ$, the bearing of Q from A taken at P.

At Q measure $\angle AQP$, the bearing of P from Q, taken at Q.

Then in $\triangle APQ$.

PQ is known.

$\angle APQ$ is known.

$\angle AQP$ is known.

∴ $\triangle APQ$ can be solved as in Case III, of § 104.

Thus AP is found and $\angle APB$ is known.

∴ $$AB = AP \sin APB$$

As a check $\angle AQB$ can be observed and AQ found as above.

Then $AB = AQ \sin AQB$.

It should be noted that the distances PB and QB can be determined if required.

Alternative method.

Instead of measuring the angles APQ, AQB, we may, by using a theodolite, measure

$\angle BPQ$ at P

and $\angle PQB$ at Q

Then in $\triangle PQB$.

PQ is known.

$\angle$s BPQ, BQP are known.

∴ $\triangle PQB$ can be solved as in Case III, § 104.

Thus PB is determined.

Then $\angle APB$ being known

$$AB = PB \tan APB$$

As a check, AB can be found by using BQ and $\angle AQB$.

109. Distance of an inaccessible object.

Suppose A (Fig. 84) to be an inaccessible object whose distance is required from an observer at P.

A distance PQ is measured in any suitable direction.

$\angle APQ$, the bearing of A with regard to PQ at P is measured.

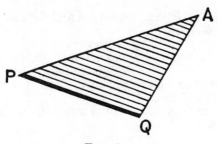

FIG. 84.

Also $\angle AQP$, the bearing of A with regard to PQ at Q is measured.

Thus in $\triangle APQ$.

PQ is known.

$\angle$s APQ, AQP are known.

$\therefore$ $\triangle APQ$ can be solved as in Case III, § 104.

Thus AP may be found and, if required, AQ.

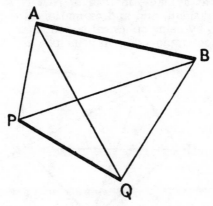

FIG. 85.

110. Distance between two visible but inaccessible objects.

Let A and B (Fig. 85) be two distant inaccessible objects.

Measure any convenient base line PQ.

At P observe $\angle$s APB, BPQ.
At Q observe $\angle$s AQP, AQB.
In $\triangle APQ$.

        PQ is known.
        $\angle$s APQ, AQP are known.

∴ $\triangle$ can be solved as in Case III, § 104, and AQ *can be found*.

Similarly $\triangle BPQ$ can be solved and QB *can be found.*
Then in $\triangle AQB$.

        AQ is known.
        QB is known.
        $\angle AQB$ is known.

∴ $\triangle AQB$ can be solved as in Case II, § 103.
Hence AB *is found.*

A check can be found by solving in a similar manner the $\triangle APB$.

III. Triangulation.

The methods employed in the last two examples are, in principle, those which are used in Triangulation. This is the name given to the method employed in surveying a district, obtaining its area, etc. In practice there are complications such as corrections for sea level and, over large districts, the fact that the earth is approximately a sphere necessitates the use of spherical trigonometry.

Over small areas, however, the error due to considering the surface as a plane, instead of part of a sphere, is, in general, very small, and approximations are obtained more readily than by using spherical trigonometry.

The method employed is, in principle, as follows:

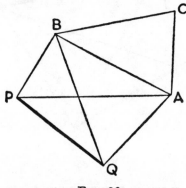

FIG. 86.

A measured distance PQ (Fig. 86), called a *base line*, is marked out with very great accuracy on suitable ground. *Then a point A is selected* and its bearings from P and Q, *i.e.* $\angle$s APQ, AQP, are observed. PQ being known, the $\triangle\,APQ$ can now be solved as in Case III and its area determined.

Next, *another point B* is selected and the angles BPA, BAP measured.

Hence, as PA has been found from $\triangle\,APQ$, $\triangle\,APB$ can be solved (Case III) and its area found.

Thus the area of the quadrilateral $PQAB$ can be found.

This can be checked by joining BQ.

The $\triangle$s BPQ, ABQ can now be solved and their areas determined.

Hence we get once more the area of the quadrilateral $PQAB$.

A new point C can now be chosen.

Using the same methods as before:

$$\triangle\,ABC \text{ can be solved.}$$

By repeating this process with other points and a network of triangles a whole district can be covered.

Not only is it essential that the base line should be measured with minute accuracy, but an extremely accurate measurement of the angles is necessary. Checks are used at every stage, such as adding the angles of a triangle to see if their sum is 180°, etc.

The instruments used, especially the theodolite, are provided with verniers and microscopic attachments to secure accurate readings.

As a further check at the end of the work, or at any convenient stage, one of the lines whose length has been found by calculation, founded on previous calculations, can be used as a base line, and the whole survey worked backwards, culminating with the calculation of the original measured base line.

112. Worked examples.

We will now consider some worked examples illustrating some of the above methods, as well as other problems solved by similar methods.

Example 1. *Two points lie due W. of a stationary balloon and are 1000 m apart. The angles of elevation at the two points are 21° 15′ and 18°. Find the height of the balloon.*

This is an example of the problem discussed under (*b*) in § 108.

In Fig. 87

$$\angle AQB = 21° 15'$$
$$\therefore \quad \angle AQP = 158° 45'$$
$$\angle APQ = 18°$$
$$\therefore \quad \angle PAQ = 3° 15'$$

$\triangle APQ$ is solved as in Case III.

$$\frac{AP}{\sin AQP} = \frac{PQ}{\sin PAQ}$$

$$\therefore \quad \frac{AP}{\sin 158° 45'} = \frac{1000}{\sin 3° 15'}$$

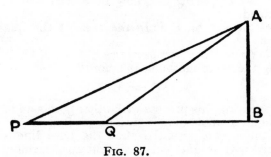

FIG. 87.

$\therefore \quad \log AP = \log 1000 + \log \sin 158° 45' - \log \sin 3° 15'$
and $\sin 158° 45' = \sin 21° 15'$ (§ 70)
whence $AP = 6395$ (see working)
also $AB = PA \sin 18°$
               $= 6395 \sin 18°$
whence $\log AB = 3 \cdot 2958$ (see working)
$\therefore \qquad \mathbf{AB = 1976 \; m}$

	Logs.
1000	3
sin 21° 15'	$\bar{1}$·5593
	2·5593
sin 3° 15'	$\bar{2}$·7535
6395	3·8058
sin 18°	$\bar{1}$·4900
1976	3·2958

Example 2. *A balloon is observed from two stations A and B at the same horizontal level, A being 1000 m north of B. At a given instant the balloon appears from A to be in a direction N. 33° 12′ E., and to have an elevation 53° 25′, while from B it appears in a direction N. 21° 27′ E. Find the height of the balloon.*

This is an example of (c) above.

In Fig. 88 PQ represents the height of the balloon at P above the ground.

$$\angle NAQ = 33° \ 12'$$
$$\angle ABQ = 21° \ 27'$$
$$\angle PAQ = 53° \ 25'$$

We first solve the $\triangle ABQ$ and so find AQ.

$$\angle BAQ = 180° - 33° \ 12' = 146° \ 48$$
$$\angle AQB = 180° - (BAQ + ABQ)$$
$$= 180° - 168° \ 15'$$
$$= 11° \ 45'$$

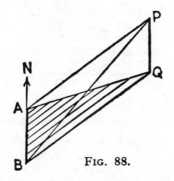

FIG. 88.

The $\triangle ABQ$ can now be solved as in Case III.

Then
$$\frac{AQ}{\sin ABQ} = \frac{AB}{\sin AQB}$$

$$\therefore \quad \frac{AQ}{\sin 21° \ 27'} = \frac{1000}{\sin 11° \ 45'}$$

$\therefore \ \log AQ = \log 1000 + \log \sin 21° \ 27' - \log \sin 11° \ 45'$
whence $AQ = 1796$ (see working)

Now $PQ = AQ \tan PAQ$
$\therefore \ PQ = 1796 \tan 53° \ 25'$
$\log PQ = \log 1796 + \log \tan 53° \ 25'$
whence $PQ = 2419$ m (see working)

	Logs.
1000	3
sin 21° 27'	1·5631
	2·5631
sin 11° 45'	1·3089
1796	3·2542
tan 53° 25'	0·1295
2419	3·3837

Example 3. *A man who wishes to find the width of a river measures along a level stretch on one bank, a line AB, 150 m long. From A he observes that a post P on the opposite bank is placed so that $\angle PAB = 51°\ 20'$, and $\angle PBA = 62°\ 12'$. What was the breadth of the river?*

In Fig. 89, AB represents the measured distance, 150 m long.

P is the post on the other side of the river.

PQ, drawn perpendicular to AB, represents the width of the river.

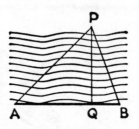

FIG. 89.

To find PQ we must first solve the $\triangle APB$.
Then knowing PA or PB we can readily find PQ.
$\triangle APB$ is solved as in *Case III*,

$$\angle PAB = 51°\ 20',\ \angle PBA = 62°\ 12'$$

$$\therefore\quad \angle APB = 180° - (51°\ 20' + 62°\ 12') = 66°\ 28'$$

$$\frac{PB}{AB} = \frac{\sin 51°\ 20'}{\sin 66°\ 28'}$$

$$\therefore\quad PB = \frac{150 \times \sin 51°\ 20'}{\sin 66°\ 28'}$$

$\log PB = \log 150 + \log \sin 51°\ 20' - \log \sin 66°\ 28'$
$\therefore\quad PB = 127 \cdot 7$ (see working)

Again $PQ = PB \sin 62°\ 12'$
$\therefore\quad \log PQ = \log 127 \cdot 7 + \log \sin 62°\ 12'$
whence $PQ = 113$ m   (see working)

This may be checked by finding PA in $\triangle PAB$ and then finding PQ as above.

	Logs.
150	2·1761
sin 51° 20′	1̄·8925
	2·0686
sin 66° 28′	1̄·9623
127·7	2·1063
sin 62° 12′	1̄·9467
113·0	2·0530

Example 4. *A and B are two ships at sea. P and Q are two stations, 1100 m apart, and approximately on the same horizontal level as A and B. At P, AB subtends an angle of 49° and BQ an angle of 31°. At Q, AB subtends an angle of 60° and AP an angle of 62°. Calculate the distance between the ships.*

Fig. 90 represents the given angles and the length PQ (not drawn to scale).

AB can be found by solving either $\triangle PAB$ or $\triangle QAB$.

To solve $\triangle PAB$ we must obtain AP and BP.

AP can be found by solving $\triangle APQ$.

BP can be found by solving $\triangle PBQ$.

In both $\triangle$s we know one side and two angles.

∴ the $\triangle$ can be solved as in Case III.

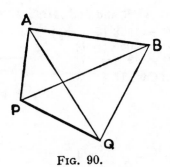

Fig. 90.

(1) To solve $\triangle APQ$ and find AP.

In $\triangle APQ$

$$\angle APQ = \angle APB + \angle BPQ$$
$$= 49° + 31° = 80°$$
$$\therefore \angle PAQ = 180° - (80° + 62°) = 38°.$$

Using the sine rule $\dfrac{AP}{PQ} = \dfrac{\sin 62°}{\sin 38°}$

$$\therefore \log AP = \log 1100 + \log \sin 62° - \log \sin 38°$$
$$= 3 \cdot 1980$$
$$\therefore AP = 1578 \text{ m} \quad (\text{see working}).$$

	Logs.
1100	3·0414
sin 62°	$\overline{1}$·9459
	2·9873
sin 38°	$\overline{1}$·7893
1578	3·1980

(2) To solve △ **BPQ** and find **BP**.

$$\angle PQB = \angle AQB + \angle AQB$$
$$= 60° + 62° = 122°.$$
$$\therefore \quad \angle PBQ = 180° - (31° + 122°) = 27°$$

Using sine rule $\dfrac{BP}{PQ} = \dfrac{\sin 122°}{\sin 27°}$

$$\therefore \quad \log BP = \log 1100 + \log \sin 122° - \log \sin 27°$$
$$= 3·3128$$
$$\therefore \quad \mathbf{BP = 2055 \text{ m}} \quad \text{(see working)}$$

	Logs.
1100	3·0414
sin 122°	$\bar{1}$·9284
	2·9698
sin 27°	$\bar{1}$·6570
2055	3·3128

(3) To solve △ **APB** and find **AB**.

We know $\quad AP = 1578 \ (= c \text{ say})$
$\qquad\qquad BP = 2055 \ (= b)$
$\qquad\qquad \angle APB = 49° \ (= A)$

$\therefore$ Solve as in Case II, § 103.

$$b = 2055$$
$$c = 1578$$

$$b + c = 3633$$
$$b - c = 477$$
$$B + C = 180° - 49°$$
$$= 131°$$

Formula used.

$$\tan \frac{B - C}{2} = \frac{b - c}{b + c} \cot \frac{A}{2}$$

Substituting

$$\tan \frac{B - C}{2} = \frac{477}{3633} \cot 24° \ 30'$$

$$\therefore \quad \log \tan \frac{B - C}{2} = \log 477 + \log \cot 24' \ 30' - \log 3633$$

$$= \bar{1}·4595$$
$$= \log \tan 16° \ 4' \quad \text{(see working)}$$

$$\therefore \quad \frac{B - C}{2} = 16° \ 4'$$

$$\qquad\qquad B - C = 32° \ 8'$$
Also $\qquad B + C = 131°$
$$\therefore \quad 2B = 163° \ 8'$$
$$\qquad\qquad B = 81° \ 34'$$
$$\qquad\qquad 2C = 98° \ 52'$$
$$\qquad\qquad C = 49° \ 26'$$
$$\therefore \quad \angle PAB = 81° \ 34'$$
$$\qquad\qquad \angle PBA = 49° \ 26'$$

	Logs.
477	2·6785
cot 24° 30'	0·3413
	3·0198
3633	3·5603
16° 4'	$\bar{1}$·4595

(4) To find **AB** use the sine rule.

$$\frac{AB}{AP} = \frac{\sin 49°}{\sin 49° 26'}$$

$$\therefore \quad AB = \frac{1578 \times \sin 49°}{\sin 49° 26'}$$

$$\therefore \quad \log AB = \log 1578 + \log \sin 49°$$
$$- \log \sin 49° 26'$$

$$= 3·1952$$

$$\therefore \quad AB = 1568 \text{ m (see working).}$$

This can be checked by solving
$\triangle AQB$ and so obtaining AQ and QB.

	Logs.
1578	3·1980
sin 49°	1̄·8778
	3·0758
sin 49° 26'	1̄·8806
1568	3·1952

Exercise 26.

1. A man observes that the angle of elevation of a tree is 32°. He walks 8 m in a direct line towards the tree and then finds that the angle of elevation is 43°. What is the height of the tree?

2. From a point Q on a horizontal plane the angle of elevation of the top of a distant mountain is 22° 18'. At a point P, 500 m further away in a direct horizontal line, the angle of elevation of the mountain is 16° 36'. Find the height of the mountain.

3. Two men stand on opposite sides of a church steeple and in the same straight line with it. They are 1·5 km apart. From one the angle of elevation of the top of the tower is 15° 30' and the other 28° 40'. Find the height of the steeple in metres.

4. A man wishes to find the breadth of a river. From a point on one bank he observes the angle of elevation of a high building on the edge of the opposite bank to be 31°. He then walks 110 m away from the river to a point in the same plane as the previous position and the building he has observed. He finds that the angle of elevation of the building is now 20° 55'. What was the breadth of the river?

5. A and B are two points on opposite sides of swampy ground. From a point P outside the swamp it is found that PA is 882 metres and PB is 1008 metres. The angle subtended at P by AB is 55° 40'. What was the distance between A and B?

6. A and B are two points 1·8 km apart on a level piece of ground along the bank of a river. P is a post on the opposite bank. It is found that $\angle PAB = 62°$ and $\angle PBA = 48°$. Find the width of the river.

7. The angle of elevation of the top of a mountain from

the bottom of a tower 180 m high is 26° 25′. From the top of the tower the angle of elevation is 25° 18′. Find the height of the mountain.

8. Two observers 5 km apart take the bearing and elevation of a balloon at the same instant. One finds that the bearing is N. 41° E. and the elevation 24°. The other finds that the bearing is N. 32° E. and the elevation 26° 37′. Calculate the height of the balloon.

9. Two landmarks A and B are observed by a man to be at the same instant in a line due east. After he has walked 4·5 km in a direction 30° north of east, A is observed to be due south while B is 38° south of east. Find the distance between A and B.

10. At a point P in a straight road PQ it is observed that two distant objects A and B are in a straight line making an angle of 35° at P with PQ. At a point C 2 km along the road from P it is observed that $\angle ACP$ is 50° and angle BCQ is 64°. What is the distance between A and B?

11. An object P is situated 345 m above a level plane. Two persons, A and B, are standing on the plane, A in a direction south-west of P and B due south of P. The angles of elevation of P as observed at A and B are 34° and 26° respectively. Find the distance between A and B.

12. P and Q are points on a straight coast line, Q being 5·3 km east of P. A ship starting from P steams 4 km in a direction 65½° N. of E.
Calculate:

(1) The distance the ship is now from the coast-line.
(2) The ship's bearing from Q.
(3) The distance of the ship from Q.

13. At a point A due south of a chimney stack, the angle of elevation of the stack is 55°. From B due west of A, such that AB = 100 m, the elevation of the stack is 33°. Find the height of the stack and its horizontal distance from A.

14. AB is a base line 0·5 km long and B is due west of A. At B a point P bears 65° 42′ north of west. The bearing of P from AB at A is 44° 15′ N. of W. How far is P from A?

15. A horizontal bridge over a river is 380 m long. From one end, A, it is observed that the angle of depression of an object, P vertically beneath the bridge, on the surface of the water is 34°. From the other end, B, the angle of depression of the object is 62°. What is the height of the bridge above the water?

16. A straight line AB, 115 m long, lies in the same

horizontal plane as the foot of a church tower PQ. The angle of elevation of the top of the tower at A is 35°. $\angle QAB$ is 62° and $\angle QBA$ is 48°. What is the height of the tower?

17. A and B are two points 1500 metres apart on a road running due west. A soldier at A observes that the bearing of an enemy's battery is 25° 48′ north of west, and at B, 31° 30′ north of west. The range of the guns in the battery is 5 km. How far can the soldier go along the road before he is within range, and what length of the road is within range?

CHAPTER X

CIRCULAR MEASURE

113. In Chapter I, when methods of measuring angles were considered, a brief reference was made to "circular measure" (§ 6 (c)), in which the unit of measurement is an angle of fixed magnitude, and not dependent upon any arbitrary division of a right angle. We now proceed to examine this in more detail.

114. Ratio of the circumference of a circle to its diameter.

The subject of "circular measure" frequently presents difficulties to the young student. In order to make it as simple as possible we shall assume, without mathematical proof, the following theorem.

The ratio of the circumference of a circle to its diameter is a fixed one for all circles.

This may be expressed in the form:

$$\frac{\text{Circumference}}{\text{diameter}} = \text{a constant.}$$

It should, of course, be noted that the ratio of the circumference of a circle to its radius is also constant and the value of the constant must be twice that of the circumference to the diameter.

115. The value of the constant ratio of circumference to diameter.

The student who is interested may obtain a fair approximation to the value of the constant by various simple experiments. For example, he may wrap a thread round a cylinder—a glass bottle will do—and so obtain the length of the circumference. He can measure the outside diameter by callipers. The ratio of circumference to diameter thus found will probably give a result somewhere about 3·14.

He can obtain a much more accurate result by the method devised by Archimedes. The perimeter of a regular polygon *inscribed* in a circle can readily be calculated. The perimeter of a corresponding *escribed* polygon can also be obtained. The mean of these two results will give an approximation to the ratio. By increasing the number of sides a still more accurate value can be obtained.

This constant is denoted by the Greek letter π (pronounced " pie ").

Hence since $$\frac{\text{circumference}}{\text{diameter}} = \pi$$

∴   circumference = π × diameter

or $c = 2\pi r$

where c = circumference and r = radius.

By methods of advanced mathematics π can be calculated to any required degree of accuracy.

To seven places

$$\pi = 3 \cdot 1415927 \ldots$$

For many purposes we take

$$\pi = 3 \cdot 1416$$

Roughly $$\pi = \frac{22}{7}.$$

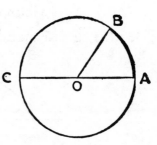

It is not possible to find any arithmetical fraction which exactly represents the value of π. Hence π is called " *incommensurable* ".

FIG. 91.

116. The unit of circular measure.

As has been stated in § 6(c) the unit of circular measure is the angle subtended at the centre of a circle by an arc whose length is equal to that of the radius.

Thus in Fig. 91 *the length of the arc AB* is equal to *r*, the radius of the circle.   The angle *AOB* is the unit by which angles are measured, and is termed a *radian*.

Definition of a radian.

A radian is the angle subtended at the centre of a circle by an arc equal in length to the radius.

Note that since

the *circumference* is π times the *diameter*
the *semicircular arc* is π times the *radius*
or arc of semicircle = π*r*.

By Theorem 17, § 18.
The angles at the centre of a circle are proportional to the arcs on which they stand.

Now in Fig. 91 the arc of the semicircle *ABC* subtends

two right angles, and the arc AB subtends 1 radian and as semicircle arc is π times arc AB

∴ angle subtended by the semicircular arc is π times the angle subtended by arc AB.

i.e.   2 right angles $= \pi$ radians
or $180° = \pi$ radians.

117. The number of degrees in a radian.

As shown above π radians $= 180°$

$$\therefore \quad 1 \text{ radian} = \frac{180°}{\pi}$$

$$= 57{\cdot}2958° \text{ approx.}$$
$$\therefore \quad 1 \text{ radian} = 57° \ 17' \ 45'' \text{ approx.}$$

118. The circular measure of any angle.

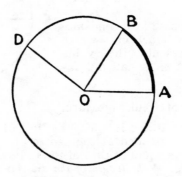

FIG. 92.

In a circle of radius r, Fig. 92, let AOD be any angle and let $\angle AOB$ represent a radian.

∴ length of arc $AB = r$.

Number of radians in $\angle AOD = \dfrac{\angle AOD}{\angle AOB}$

∴ By Theorem 17 quoted above

$$\frac{\angle AOD}{\angle AOB} = \frac{\text{arc } AD}{\text{arc } AB}$$

If $\theta =$ number of radians in $\angle AOD$

then $\theta = \dfrac{\text{arc } \boldsymbol{AD}}{r}.$

119. To convert degrees to radians.

Since $\qquad 180° = \pi$ radians

$$\therefore \quad 1° = \frac{\pi}{180} \text{ radians}$$

and $\qquad \theta° = \left(\theta \times \frac{\pi}{180}\right)$ **radians.**

120. To find the length of an arc.

Let $\qquad a =$ length of arc

and $\qquad \theta =$ number of radians in angle.

Then as shown in § 118

$$\frac{\text{arc}}{\text{radius}} = \text{number of radians in the angle the arc subtends.}$$

$$\therefore \quad \frac{a}{r} = \theta \quad (\S\ 118)$$

and $\qquad \boldsymbol{a = r\theta.}$

121. In more advanced mathematics, circular measure is always employed except in cases in which, for practical purposes, we require to use degrees. Consequently when we speak of an angle θ, it is generally understood that we are speaking of θ radians. Thus when referring to π radians, the equivalent of two right angles, we commonly speak of the angle π. Hence we have the double use of the symbol:

(1) As the constant ratio of the circumference of a circle to its diameter;

(2) As short for π radians, *i.e.* the equivalent of 180°.

In accordance with this use of π, angles are frequently expressed as multiples or fractions of it.

Thus
$$2\pi = 360°$$
$$\frac{\pi}{2} = 90°$$
$$\frac{\pi}{4} = 45°$$
$$\frac{\pi}{3} = 60°$$
$$\frac{\pi}{6} = 30°$$

π is not usually evaluated in such cases, except for some special purpose.

Exercise 27

1. What is the number of degrees in each of the following angles expressed in radians: $\dfrac{\pi}{3}, \dfrac{\pi}{12}, \dfrac{3\pi}{2}, \dfrac{2\pi}{3}, \dfrac{3\pi}{4}$?

2. Write down from the tables the following ratios:

    (a) $\sin \dfrac{\pi}{5}$ (b) $\cos \dfrac{\pi}{8}$.   (c) $\sin \dfrac{\pi}{10}$.

    (d) $\cos \dfrac{3\pi}{8}$.   (e) $\sin \left(\dfrac{\pi}{3} + \dfrac{\pi}{4}\right)$.

3. Express in radians the angles subtended by the following arcs:

    (a) arc = 11·4 mm, radius = 2·4 mm.
    (b) arc = 5·6 cm,   radius = 2·2 cm.

4. Express the following angles in degrees and minutes:

    (a) 0·234 radian.   (b) 1·56 radian.

5. Express the following angles in radians, using fractions of π:

    (a) 15°.   (b) 72°.   (c) 66°.   (d) 105°.

6. Find the length of the arc in each of the following cases:

    (1) r = 2·3 cm, θ = 2·54 radians.
    (2) r = 12·5 m, θ = 1·4 radians.

7. A circular arc 154 cm long and the radius of the arc is 252 cm. What is the angle subtended at the centre of the circle, in radians and degrees?

8. Express a right angle in radians, not using a multiple of π.

9. The angles of a triangle are in the ratio of 3 : 4 : 5. Express them in radians.

CHAPTER XI

TRIGONOMETRICAL RATIOS OF ANGLES OF ANY MAGNITUDE

122. In Chapter III we dealt with the trigonometrical ratios of acute angles, *i.e.* angles in the first quadrant. In Chapter V the definitions of these ratios were extended to obtuse angles, or angles in the second quadrant. But in mathematics we generalise and consequently in this chapter we proceed to consider the ratios of angles of any magnitude.

In § 5, Chapter I, an angle was defined by the rotation of a straight line from a fixed position and round a fixed centre, and there was no limitation as to the amount of rotation. The rotating line may describe any angle up to 360° or one complete rotation, and may then proceed to two, three, four—to any number of complete rotations in addition to the rotation made initially.

123. Angles in the third and fourth quadrants.

We will first deal with angles in the third and fourth quadrants, and thus include all those angles which are less than 360° or a complete rotation.

Before proceeding to the work which follows the student is advised to revise § 68, in Chapter V, dealing with positive and negative lines.

In § 70 it was shown that the ratios of angles in the second quadrant were defined in the same fundamental method as those of angles in the first quadrant, the only difference being that in obtaining the values of the ratios we have to take into consideration the *signs of the lines employed*, *i.e.* whether they are positive or negative.

It will now be seen that, with the same attention to the signs of the lines, the same definitions of the trigonometrical ratios will apply, whatever the quadrant in which the angle occurs.

In Fig. 93 there are shown in separate diagrams, angles in the four quadrants. In each case from a point P on the rotating line a perpendicular PQ is drawn to the fixed line OX, produced in the cases of the second and third quadrants.

Thus we have formed, in each case, a triangle OPQ, using

the sides of which we obtain, in each quadrant, the ratios as follows, denoting $\angle AOP$ by θ.

Then, in each quadrant

$$\sin \theta = \frac{PQ}{OP}$$

$$\cos \theta = \frac{OQ}{OP}$$

$$\tan \theta = \frac{PQ}{OQ}$$

We now consider the signs of these lines in each quadrant.

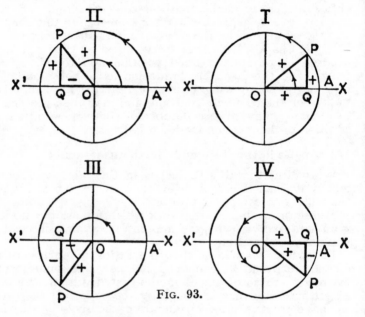

Fig. 93.

(1) In the first quadrant.

All the lines are +ve.
∴ All the ratios are +ve.

(2) In the second quadrant.

OQ is −ve
∴ $\sin \theta$ is +ve
$\cos \theta$ is −ve
$\tan \theta$ is −ve

(3) In the third quadrant.

$$OQ \text{ and } PQ \text{ are } -ve$$

∴ $\sin \theta$ is $-ve$

$\cos \theta$ is $-ve$

$\tan \theta$ is $+ve$

(4) In the fourth quadrant.

$$PQ \text{ is } -ve$$

∴ $\sin \theta$ is $-ve$

$\cos \theta$ is $+ve$

$\tan \theta$ is $-ve$

Note.—The cosecant, secant and tangent will, of course, have the same signs as their reciprocals. These results may be summarised as follows:

	Quadrant II	Quadrant I	
sine +	sin, +ve	sin, +ve	
	cos, −ve	cos, +ve	all +
	tan, −ve	tan, +ve	

	Quadrant III	Quadrant IV	
tan +	sin, −ve	sin, −ve	
	cos, −ve	cos, +ve	cos +
	tan, +ve	tan, −ve	

124. Variations in the sine of an angle between 0° and 360°.

These have previously been considered for angles in the first and second quadrants. Summarising these for completeness, we will proceed to examine the changes in the third and fourth quadrants.

Construct a circle of unit radius (Fig. 94) and centre O, Take on the circumference of this a series of points P_1, P_2. P_3 . . . and draw perpendiculars to the fixed line XOX'. Then the radius being of unit length, these perpendiculars, in the scale in which OA represents unity, will represent the sines of the corresponding angles.

By observing the changes in the lengths of these perpendiculars we can see, throughout the four quadrants, the changes in the value of the sine.

In quadrant I

$\sin \theta$ is $+ve$ and increasing from 0 to 1.

In quadrant II

$\sin \theta$ is $+ve$ and decreasing from 1 to 0.

In quadrant III

sin θ is −ve.

Now the actual lengths of the perpendiculars is increasing, but as they are −ve, the value of the sine is actually decreasing, and at 270° is equal to − 1.

∴   The sine decreases in this quadrant from 0 to − 1.

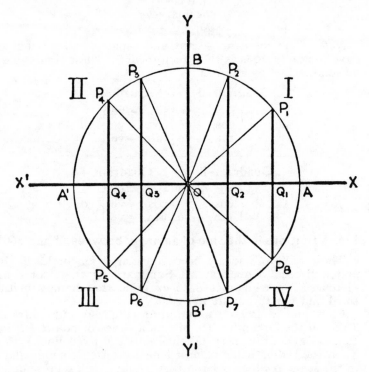

Fig. 94.

In quadrant IV

sin θ is −ve.

The lengths of the perpendiculars are decreasing, but as they are −ve, their values are increasing and at 360° the sine is equal to sin 0° and is therefore zero.

∴   sin θ is increasing from − 1 to 0.

125. Graphs of sin θ and cosec θ.

By using the values of sines obtained in the method shown above (Fig. 94) or by taking the values of sines from the tables, a graph of the sine between 0° and 360° can now be drawn. It is shown in Fig. 95, together with that of cosec θ (dotted line) the changes in which through the four quadrants can be deduced from those of the sine. The student should compare the two graphs, their signs, their maximum and minimum values, etc.

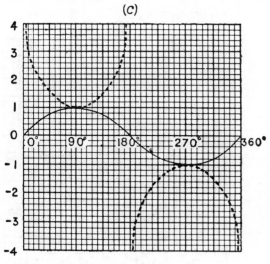

Fig. 95.

Graphs of *sin* θ and *cosec* θ.

(*Cosec* θ is dotted.)

126. Variations in the cosine of an angle between 0° and 360°.

If the student will refer to Fig. 94, he will see that the distances intercepted on the fixed line by the perpendiculars from P_1, P_2 . . ., viz. OQ_1, OQ_2 . . . will represent, in the scale in which OA represents unity, the cosines of the corresponding angles. Examining these we see

(1) In quadrant I.

The cosine is +ve and decreases from 1 to 0.

(2) **In quadrant II.**

The cosine is always −ve and decreases from 0 to − 1.

(3) **In quadrant III.**

The cosine is −ve and always increasing from − 1 to 0 and cos 270° = 0.

(4) **In quadrant IV.**

The cosine is +ve and always increasing from 0 to + 1 since cos 360° = cos 0° = 1.

(a)

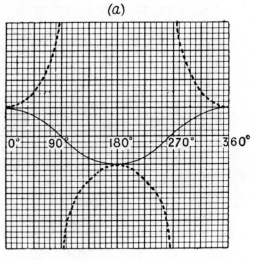

FIG. 96.

Graphs of *cos* θ and *sec* θ (dotted curve).

127. Graphs of cos θ and sec θ.

In Fig. 96 is shown the graph of cos θ, which can be drawn as directed for the sine in § 125. The curve of its reciprocal, sec θ, is also shown by the dotted curve. These two curves should be compared by the student.

128. Variations in the tangent between 0° and 360°.

The changes in the value of tan θ between 0° and 360° can be seen in Fig. 97, which is an extension of Fig. 39.

The circle is drawn with unit radius.

From A and A' tangents are drawn to the circle and at right angles to XOX'.

Considering any angle such as AOP_1,

$$\tan AOP_1 = \frac{P_1A}{OA} = \frac{P_1A}{1} = P_1A$$

Consequently P_1A, P_2A, P_3A', P_4A' . . . represent the *numerical* value of the tangent of the corresponding angle.

But account must be taken of the sign.

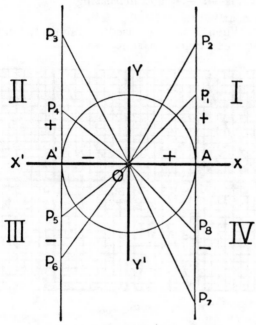

FIG. 97.

In quadrants II and III, the denominator of the ratio is -1 in numerical value, while in quadrants III and IV the numerator of the fraction is $-$ve.

Consequently the tangent is $+$ve in quadrants I and III and $-$ve in quadrants II and IV.

Considering a particular angle, viz. the $\angle A'OP_5$ in the quadrant III

$$\tan A'OP_5 = \frac{-P_5A'}{-OA'}$$

$\therefore$ $\tan \theta$ is $+$ve and is represented numerically by P_5A'.

From such observations of the varying values of tan θ the changes between 0° and 360° can be determined as follows:

(1) In quadrant I

tan θ is always +ve and increasing.

It is 0 at 0° and ──→ ∞ at 90°.

(2) In quadrant II

tan θ is always −ve and increasing

from − ∞ at 90° to 0 at 180°.

Note.—When θ has increased an infinitely small amount above 90°, the tangent becomes −ve.

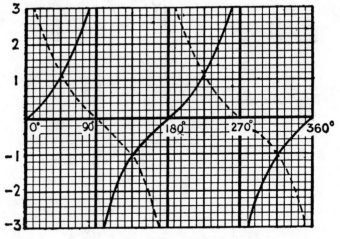

Fig. 98.

Graph of tan θ and cot θ (dotted line)

(3) In quadrant III

tan θ is always +ve and increasing.

At 180° the tangent is 0 and at 270° tan θ ──→ ∞ .

(4) In quadrant IV

tan θ is always −ve and increasing

from − ∞ at 270° to 0 at 360°.

129. Graphs of tan θ and cot θ.

In Fig. 98 are shown the graphs of tan θ and cot θ (dotted curve) for values of angles between 0° and 360°.

130. Ratios of angles greater than 360°.

Let $\angle AOP$ (Fig. 99) be any angle, θ, which has been formed by rotation in an anti-clockwise or positive direction from the position OA.

Suppose now that the rotating line continues to rotate in the same direction for a complete rotation or 360° from OP so that it arrives in the same position, OP, as before. The total amount of rotation from OA is now 360° + θ or $(2\pi + \theta)$ radians.

Clearly the trigonometrical ratios of this new angle $2\pi + \theta$ must be the same as θ, so that sin $(2\pi + \theta) = \sin \theta$, and so for the other ratios.

Similarly if further complete rotations were made so that angles were formed such as $4\pi + \theta$, $6\pi + \theta$, etc., it is evident that the trigonometrical ratios of these angles will be the same as those of θ.

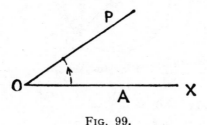

Fig. 99.

Turning again to Fig. 99 it is also evident that if a complete rotation were made in a clockwise, *i.e.* negative, direction, from the position OP, we should have the angle $- 2\pi + \theta$. The trigonometrical ratios of this angle, and also such angles as $- 4\pi + \theta$, $- 6\pi + \theta$, will be the same as those of θ.

All such angles can be included in the general formula

$$2n\pi + \theta$$

where " n " is any integer, positive or negative.

Referring to the graphs of the ratios in Figs. 95, 96 and 98, it is clear that when the angle is increased by successive complete rotations, the curves as shown, will be repeated either in a positive or a negative direction, and this can be done to an infinite extent.

Each of the ratios is called a " periodic function " of the angle, because the values of the ratio are repeated at intervals of 2π radians or 360°, which is called the period of the function.

131. Trigonometrical ratio of − θ.

In Fig. 100 let the rotating line *OA* rotate in a clockwise, *i.e.* negative, direction to form the angle *AOP*. This will be a negative angle. Let it be represented by − θ.

Let the angle *AOP′* be formed by rotation in an anticlockwise *i.e.* +ve direction and let it be equal to θ.

Then the straight line *P′MP* completes two triangles.

OMP and *OMP′*

These triangles are congruent (Theorem 7, § 13) and the angles *OMP*, *OMP′* are equal and ∴ right angles.

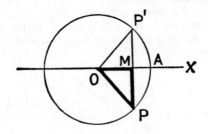

Fɪɢ. 100.

Then $\qquad \sin(-\theta) = \dfrac{PM}{OP} = -\dfrac{P'M}{OP}$

but $\qquad\qquad \dfrac{PM}{OP} = \sin\theta$

∴ $\quad \sin(-\theta) = -\sin\theta$

Similarly $\quad \cos(-\theta) = \dfrac{OM}{OP} = \dfrac{OM}{OP'} = \cos\theta$

Similarly $\qquad\qquad \tan(-\theta) = -\tan\theta$

Collecting these results,

$$\sin(-\theta) = -\sin\theta$$
$$\cos(-\theta) = \cos\theta$$
$$\tan(-\theta) = -\tan\theta$$

By these results the student will be able to construct the curves of sin θ, cos θ and tan θ for −ve angles. He will see that the curves for −ve angles will be repeated in the opposite direction.

132. To compare the trigonometrical ratios of θ and 180° + θ.

Note.—If θ is an acute angle, then 180° + θ or π + θ is an angle in the third quadrant.

In Fig. 101 with the usual construction let $\angle POQ$ be any acute angle, θ.

Let PO be produced to meet the circle again in P'.

Draw PQ and $P'Q'$ perpendicular to XOX'.

Then $\qquad \angle P'OQ' = \angle POQ = \theta$ (Theorem 1, § 8)

and $\qquad \angle AOP' = 180° + \theta$.

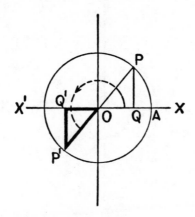

FIG. 101.

The $\triangle$s POQ, $P'OQ'$ are congruent

$$P'Q' = -PQ$$
and
$$OQ' = -OQ$$

Now $\qquad \sin \theta = \dfrac{PQ}{OP}$

and $\qquad \sin (180° + \theta) = \sin AOP'$

$$= \frac{P'Q'}{OP'} = \frac{-PQ}{OP} = -\sin \theta$$

$$\therefore \quad \sin \theta = -\sin (180° + \theta)$$
similarly $\qquad \cos \theta = -\cos (180° + \theta)$
and $\qquad \tan \theta = \tan (180° + \theta)$

133. To compare the ratios of θ and $360° - \theta$.

Note.—If θ is an acute angle, then $360° - \theta$ is an angle in the fourth quadrant.

In Fig. 102 if the acute angle AOP represents θ then the re-entrant angle AOP, shown by the dotted line represents $360° - \theta$.

The trigonometrical ratios of this angle may be obtained

from the sides of the $\triangle OMP$ in the usual way and will be the same as those of $-\theta$ (see § 131).

∴   using the results of § 131 we have

$$\sin (360° - \theta) = - \sin \theta$$
$$\cos (360° - \theta) = \cos \theta$$
$$\tan (360° - \theta) = - \tan \theta$$

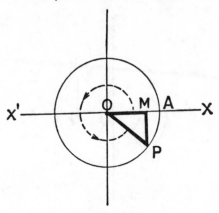

FIG. 102.

134. It will be convenient for future reference to collect some of the results obtained in this chapter, as follows

$$\sin \theta = \sin (\pi - \theta) = - \sin (\pi + \theta) = - \sin (2\pi - \theta)$$
$$= - \sin (- \theta)$$
$$\cos \theta = - \cos (\pi - \theta) = - \cos (\pi + \theta) = \cos (2\pi - \theta$$
$$= \cos (- \theta$$
$$\tan \theta = - \tan (\pi - \theta) = \tan (\pi + \theta) = - \tan (2\pi - \theta)$$
$$= - \tan (- \theta)$$

135. It is now possible, by use of the above results and using the tables of ratios for acute angles, to write down the trigonometrical ratios of angles of any magnitude.

A few examples are given to illustrate the method to be employed.

Example I.   *Find the value of sin 245°.*

We first note that this angle is in the third quadrant
∴   its sine must be negative.
Next, by using the form of $(180° + \theta)$

$$\sin 245° = \sin (180° + 65°)$$

Thus we can use the appropriate formula of § 134, viz.

$$\sin \theta = - \sin (\pi + \theta)$$

Consequently
$$\sin (180° + 65°) = -\sin 65°$$
$$= -0\cdot9063.$$

Example 2. *Find the value of cos 325°.*

This angle is in the fourth quadrant and so we use the formulae for values of $360° - \theta$ (see § 133).

In this quadrant the cosine is always $+$ve

$$\cos 325° = \cos (360° - 35°)$$
$$= \cos 35° \qquad \text{(§ 133)}$$
$$= 0\cdot8192.$$

Example 3. *Find the value of tan 392°.*

This angle is greater than 360° or one whole revolution.

$$\therefore \quad \tan 392° = \tan (360° + 32°)$$
$$= \tan 32°$$
$$= 0\cdot6249.$$

Example 4. *Find the value of sec 253°.*

This angle is in the third quadrant.

$\therefore$ we use the formula connected with $(\pi + \theta)$ (see § 132). Also in this quadrant the cosine, the reciprocal of the secant is $-$ve.

$$\sec 253° = \sec (180° + 73°)$$
$$= -\sec 73°$$
$$= -3\cdot4203.$$

Exercise 28

1. Find the sine, cosine and tangent of each of the following angles:

    (a) 257°.   (b) 201° 13′.
    (c) 315° 20′.   (d) 343° 8′.

2. Find the values of:

    (a) $\sin (-51°.)$ (b) $\cos (-42°).$
    (c) $\sin (-138°).$ (d) $\cos (-256°).$

3. Find the values of:

    (a) cosec 251°.   (b) sec 300°.
    (c) cot 321°.   (d) sec 235°.

4. Find the values of:

    (a) $\sin (\pi + 57°).$ (b) $\cos (2\pi - 42°).$
    (c) $\tan (2\pi + 52°).$ (d) $\sin (4\pi + 36°).$

136. To find the angles which have given trigonometrical ratios.

(a) **To find all the angles which have a given sine (or cosecant).**

We have already seen in § 73 that corresponding to a given sine there are two angles, θ and $180° - \theta$, where θ is the acute angle whose sine is given in the tables.   Having now considered angles of any magnitude it becomes necessary to discover what other angles have the given sine.

An examination of the graph of sin θ in Fig. 95 shows that only two of the angles less than 360° have a given sine, whether it be positive or negative, the two already mentioned above if the sine is +ve, and two in the third and fourth quadrants if it be −ve.

But the curve may extend to an indefinite extent for angles greater than 360°, and for negative angles, and every section corresponding to each additional 360°, positive or negative, will be similar to that shown.   Therefore it follows that there will be an infinite number of other angles, two in each section which have the given sine.   These will occur at intervals of 2π radians from those in the first quadrant. There will thus be two sets of such angles.

(1) θ, $2\pi + \theta$, $4\pi + \theta$, . . .
(2) $\pi - \theta$, $3\pi - \theta$, $5\pi - \theta$, . . .

These two sets include all the angles which have the given sines.   They can be summarised as follows:

(1) (any *even* multiple of π) $+ \theta$.
(2) (any *odd* multiple of π) $- \theta$.

These can be combined together in one formula as follows:

Let *n* be any integer, positive or negative.
Then sets (1) and (2) are contained in

$$n\pi + (-1)^n \theta$$

The introduction of $(-1)^n$ is a device which ensures that when *n is even*, *i.e.* we have an even multiple of π, $(-1)^n = 1$ and the formula becomes $n\pi + \theta$.   When *n is odd* $(-1)^n = -1$ and the formula becomes $n\pi - \theta$.

∴ the general formula for all angles which have a given sine is

$$n\pi + (-1)^n \theta$$

where *n* is any integer +ve or −ve, and θ is the smallest angle having the given sine.
The same formula will clearly hold also for the cosecant.

(*b*) **To find all the angles which have a given cosine (or secant).**

Examining the graph of cos θ (Fig. 96), it is seen that there are two angles between 0° and 360° which have a given cosine which is +ve, one in the first quadrant and one in the fourth.   If the given cosine is −ve, the two angles lie in the second and third quadrants. These two angles are expressed by θ and $360° - \theta$.

or $\qquad\qquad \theta$ and $2\pi - \theta$ in radians.

As in the case of the sine for angles greater than 360° or for negative angles, there will be two angles with the given sine in the section corresponding to each additional 360°.

There will therefore be two sets:

(1) $\theta,\ 2\pi + \theta,\ 4\pi + \theta,\ \ldots$
(2) $2\pi - \theta,\ 4\pi - \theta,\ 6\pi - \theta,\ \ldots$

These can be combined in one set, viz.:

$$\text{(any even multiple of } \pi) \pm \theta$$

or if n be any integer, positive on negative, this can be expressed by

$$2n\pi \pm \theta.$$

∴ the general formula for all angles with a given cosine is:

$$2n\pi \pm \theta.$$

The formula for the secant will be the same.

(c) **To find all the angles which have a given tangent (or cotangent).**

An examination of the graph of $\tan \theta$ (Fig. 98), shows that there are two angles less than 360° which have the same tangent, viz.:

$$\theta \text{ and } 180° + \theta$$
or $\qquad \theta$ and $\ \ \pi + \theta$

As before, there will be other angles at intervals of 2π which will have the same tangent. Thus there will be two sets, viz.:

$$\theta,\ 2\pi + \theta,\ 4\pi + \theta,\ \ldots$$
$$\pi + \theta,\ 3\pi + \theta,\ 5\pi + \theta,\ \ldots$$

Combining these it is clear that all are included in the general formula

$$\text{(any multiple of } \pi) + \theta$$

∴ If n be any integer, positive or negative,

The general formula for all angles with a given tangent is
$$n\pi + \theta$$

The same formula holds for the cotangent.

Exercises which involve the use of these formulae will occur in the next chapter.

CHAPTER XII

TRIGONOMETRICAL EQUATIONS

137. TRIGONOMETRICAL equations are those in which the unknown quantities, whose values we require, are the trigonometrical ratios of angles. The angles themselves can be determined when the values of the ratios are known.

The actual form which the answer will take depends on whether we require only the smallest angle corresponding to the ratio, which will be obtained from the tables, or whether we want to include some or all of those other angles which, as we have seen in the previous chapter, have the same ratio.

This can be shown in a very simple example.

Example. *Solve the equation* $2 \cos \theta = 0.842$.

(1) **The smallest angle only may be required.**

Since
$$2 \cos \theta = 0.842$$
$$\cos \theta = 0.421$$
From the tables $\qquad \theta = 65° 6'$.

(2) **The angles between 0° and 360° which satisfy the equation may be required.**

As we have seen in § 136(*b*) there is only one other such angle, in the fourth quadrant.

It is given by $\qquad 2\pi - \theta$ or $360° - \theta$

∴ This angle $= 360° - 65° 6' = 294° 54'$.

∴ The two solutions are $65° 6'$ and $294° 54'$.

(3) **A general expression for all angles which satisfy the equation may be required.**

In this case one of the formulae obtained in the previous chapter will be used.

Thus in § 136(*b*) all angles with a given cosine are included in the formula
$$2n\pi \pm \theta$$

In this example $\theta = 65° 6'$.

∴ The solution is $2n\pi \pm \cos^{-1} 0.421$.

The inverse notation (see § 74) is used to avoid the incongruity of part of the answer $2n\pi$ being in radians, and the other in degrees.

138. Some of the different types of equations will now be considered.

(*a*) **Equations which involve only one ratio.**

The example considered in the previous paragraph is the simplest form of this type.   Very little manipulation is, as a rule, required unless the equation is quadratic in form.

Example.   *Solve the equation*

$$6 \sin^2 \theta - 7 \sin \theta + 2 = 0$$

for values of θ *between* $0°$ *and* $360°$.

Factorising

$$(3 \sin \theta - 2)(2 \sin \theta - 1) = 0$$

whence $3 \sin \theta - 2 = 0$ (1)

or $2 \sin \theta - 1 = 0$ (2)

From (1) $\sin \theta = \frac{2}{3} = 0.6667$

∴   from the tables

$$\theta = 41° 49'.$$

The only other angle less than $360°$ with this sine is

$$180° - \theta = 138° 11'$$

From (2) $\sin \theta = \frac{1}{2}$

∴ $\theta = 30°.$

and the other angle with this sine is $180° - 30°$

$$= 150°$$

∴   the complete solution is

$$41° 49', 138° 11', 30°, 150°.$$

Note.—If one of the values of $\sin \theta$ or $\cos \theta$ obtained in an equation is numerically greater than unity, such a root must be discarded as impossible.   Similarly values of the secant and cosecant less than unity are impossible solutions from this point of view.

(*c*) **Equations containing more than one ratio of the angle.**

Manipulation is necessary to replace one of the ratios by its equivalent in terms of the other.   To effect this we must use an appropriate formula connected with the ratios such as were proved in Chapter IV.

Example I.   *Obtain a complete solution of the equation*

$$3 \sin \theta = 2 \cos^2 \theta.$$

The best plan here is to change $\cos^2 \theta$ into its equivalent value of $\sin \theta$.   This can be done by the formula

$$\sin^2 \theta + \cos^2 \theta = 1$$

whence $\cos^2 \theta = 1 - \sin^2 \theta.$

Substituting in the above equation
$$3 \sin \theta = 2(1 - \sin^2 \theta)$$
$$\therefore \quad 2 \sin^2 \theta + 3 \sin \theta - 2 = 0$$

Factorising,
$$(2 \sin \theta - 1)(\sin \theta + 2) = 0$$

whence $\sin \theta + 2 = 0$ (1)
or $2 \sin \theta - 1 = 0$ (2)

From (1) $\sin \theta = -2$

This is impossible, and therefore does not provide a solution of the given equation.

From (2) $2 \sin \theta = 1$
$$\therefore \quad \sin \theta = \tfrac{1}{2}$$

The smallest angle with this sine is $30°$ or $\dfrac{\pi}{6}$ radians.

Using the general formula for all angles with a given sine, viz.:
$$n\pi + (-1)^n \theta$$

The general solution of the equation is
$$\theta = n\pi + (-1)^n \frac{\pi}{6}$$

Example 2. *Solve the equation*
$$sin\ 2\theta = cos^2\ \theta$$

giving the values of θ between $0°$ and $360°$ which satisfy the equation.

Since $\sin 2\theta = 2 \sin \theta \cos \theta$ (see § 83)
$$\therefore \quad 2 \sin \theta \cos \theta = \cos^2 \theta$$
Hence $\cos \theta = 0$ (1)
or $2 \sin \theta = \cos \theta$ (2)
From (1) $\theta = 90°$ or $270°$

From (2) $2 \sin \theta = \cos \theta$
$$\therefore \quad 2 \tan \theta = 1$$
and $\tan \theta = 0{\cdot}5$
whence $\theta = 26° \; 34'$
Also $\tan \theta = \tan (180° + \theta)$ (see § 132)

$\therefore$ The other angle less than $360°$ with this tangent is
$$180° + 26° \; 34'$$
$$= 206° \; 34'$$

$\therefore$ The solution is $\theta = 26° \; 34'$ or $206° \; 34'$.

$\therefore$ The required solution is $\theta = 90°$, $270°$, $26° \; 34'$ or $206° \; 34'$.

139. Equations of the form

$$a \cos \theta + b \sin \theta = c.$$

where a, b, c are known constants, are important in electrical work and other applications of trigonometry.

This could be solved by using the substitution

$$\cos \theta = \sqrt{1 - \sin^2 \theta}$$

but the introduction of the square root is not satisfactory. We can obtain a solution more readily by the following device.

Since a and b are known it is always possible to find an angle α, such that

$$\tan \alpha = \frac{a}{b}$$

as the tangent is capable of having any value (see graph, Fig. 98).

Let ABC (Fig. 103) be a right-angled $\triangle$ in which the sides congaining the right angle are a and b units in length.

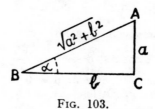

Fig. 103.

Then
$$\tan ABC = \frac{a}{b}$$
$$\therefore \quad \angle ABC = \alpha$$

By the Theorem of Pythagoras:

$$AB = \sqrt{a^2 + b^2}$$

and
$$\frac{a}{\sqrt{a^2 + b^2}} = \sin \alpha$$

$$\frac{b}{\sqrt{a^2 + b^2}} = \cos \alpha$$

$\therefore$ in the equation

$$a \cos \theta + b \sin \theta = c$$

Divide throughout by $\sqrt{a^2 + b^2}$

$$\therefore \quad \frac{a}{\sqrt{a^2 + b^2}} \cos \theta + \frac{b}{\sqrt{a^2 + b^2}} \sin \theta = \frac{c}{\sqrt{a^2 + b^2}}$$

$$\therefore \quad \sin \alpha \cos \theta + \cos \alpha \sin \theta = \frac{c}{\sqrt{a^2 + b^2}}$$

$$\therefore \quad \sin (\theta + \alpha) = \frac{c}{\sqrt{a^2 + b^2}} \qquad \text{(see § 80, No. 1)}$$

Now $\dfrac{c}{\sqrt{a^2 + b^2}}$ can be evaluated, since a, b, c are known and provided it is less than unity it is the sine of some angle, say β.

$$\therefore \quad \theta + \alpha = \beta$$
and
$$\theta = \beta - \alpha$$

Thus the least value of θ is determined.

Example. *Solve the equation*

$$3 \cos \theta + 4 \sin \theta = 3 \cdot 5$$

In this case $a = 3$, $b = 4$,

$$\therefore \quad \sqrt{a^2 + b^2} = \sqrt{9 + 16} = 5.$$

Thus $\tan \alpha = \frac{3}{4}$, $\sin \alpha = \frac{3}{5}$, $\cos \alpha = \frac{4}{5}$ and $\alpha = 36° \ 52'$ (from the tables).

$\therefore$ Dividing the given equation by 5

$$\tfrac{3}{5} \cos \theta + \tfrac{4}{5} \sin \theta = \frac{3 \cdot 5}{5}$$

$$\therefore \quad \sin \alpha \cos \theta + \cos \alpha \sin \theta = 0 \cdot 7$$
$$\therefore \quad \sin (\theta + \alpha) = 0 \cdot 7$$

But the angle whose sine is $0 \cdot 7$ is $44° \ 25'$.

$$\therefore \quad \theta + \alpha = 44° \ 25'$$
or
$$\theta + 36° \ 52' = 44° \ 25'$$
$$\therefore \quad \theta = 44° \ 25' - 36° \ 52'$$
$$= 7° \ 33'.$$

139. Variations of $a \cos \theta + b \sin \theta$.

This expression is an important one in its application, and the graphical representations of its variation may have to be studied by some students. The variations of the expression may be best studied by using, in a modified form, the device employed above.

By means of the reasoning given in the previous paragraph, the expression can be written in the form

$$\sqrt{a^2 + b^2} \{\sin (\theta + \alpha)\}$$

By assigning different values to θ, the only variable in the expression, the variations can be studied and a graph constructed.

Exercise 29

1. Find the angles less than 360° which satisfy the following equations:

    (1) $\sin \theta = 0.8910$.   (2) $\cos \theta = 0.4179$.
    (3) $2 \tan \theta = 0.7$.   (4) $\sec \theta = 2.375$.

2. Find the angles less than 360° which satisfy the following equations:

    (1) $4 \cos 2\theta - 3 = 0$.   (2) $3 \sin 2\theta = 1.8$.

3. Find the angles less than 360° which satisfy the following equations:

    (1) $6 \sin \theta = \tan \theta$.   (2) $4 \cos \theta = 3 \tan \theta$.
    (3) $3 \cos^2 \theta + 5 \sin^2 \theta = 4$.   (4) $4 \cos \theta = 3 \sec \theta$.

4. Find the angles less than 360° which satisfy the following equations:

    (1) $2 \tan^2 \theta - 3 \tan \theta + 1 = 0$.
    (2) $5 \tan^2 \theta - \sec^2 \theta = 11$.
    (3) $4 \sin^2 \theta - 3 \cos \theta = 1.5$.
    (4) $\sin \theta + \sin^2 \theta = 0$.

5. Find general formulae for the angles which satisfy the following equations:

    (1) $2 \cos \theta - 0.6578 = 0$.
    (2) $\frac{1}{2} \sin 2\theta = 0.3174$.
    (3) $\cos 2\theta + \sin \theta = 1$.
    (4) $\tan \theta + \cot \theta = 4$.

6. Find the smallest angles which satisfy the equations:

    (1) $\sin \theta + \cos \theta = 1.2$.
    (2) $\sin \theta - \cos \theta = 0.2$.
    (3) $2 \cos \theta + \sin \theta = 2.1$.
    (4) $4 \cos \theta + 3 \sin \theta = 5$.

SUMMARY OF FORMULAE

1. Complementary angles.

$$\sin \theta = \cos (90° - \theta)$$
$$\cos \theta = \sin (90° - \theta)$$
$$\tan \theta = \cot (90° - \theta)$$

2. Supplementary angles.

$$\sin \theta = \sin (180° - \theta)$$
$$\cos \theta = -\cos (180° - \theta)$$
$$\tan \theta = -\tan (180° - \theta)$$

3. Relations between the ratios.

$$\tan \theta = \frac{\sin \theta}{\cos \theta}$$
$$\sin^2 \theta + \cos^2 \theta = 1$$
$$\tan^2 \theta + 1 = \sec^2 \theta$$
$$\cot^2 \theta + 1 = \operatorname{cosec}^2 \theta$$

4. Compound angles.

$$\sin (A + B) = \sin A \cos B + \cos A \sin B$$
$$\sin (A - B) = \sin A \cos B - \cos A \sin B$$
$$\cos (A + B) = \cos A \cos B - \sin A \sin B$$
$$\cos (A - B) = \cos A \cos B + \sin A \sin B$$
$$\tan (A + B) = \frac{\tan A + \tan B}{1 - \tan A \tan B}$$
$$\tan (A - B) = \frac{\tan A - \tan B}{1 + \tan A \tan B}$$
$$\sin P + \sin Q = 2 \sin \frac{P+Q}{2} \cos \frac{P-Q}{2}$$
$$\sin P - \sin Q = 2 \cos \frac{P+Q}{2} \sin \frac{P-Q}{2}$$
$$\cos P + \cos Q = 2 \cos \frac{P+Q}{2} \cos \frac{P-Q}{2}$$
$$\cos Q - \cos P = 2 \sin \frac{P+Q}{2} \sin \frac{P-Q}{2}$$

5. Multiple angles.

$$\sin 2\theta = 2 \sin \theta \cos \theta$$
$$\cos 2\theta = \cos^2 \theta - \sin^2 \theta$$
$$= 2 \cos^2 \theta - 1$$
$$= 1 - 2 \sin^2 \theta$$

or

$$\cos^2 \theta = \tfrac{1}{2}(1 + \cos 2\theta)$$
$$\sin^2 \theta = \tfrac{1}{2}(1 - \cos 2\theta)$$
$$\tan 2\theta = \frac{2 \tan \theta}{1 - \tan^2 \theta}$$

6. Solutions of a triangle.

   Case I.   Three sides known.

      1. $\cos A = \dfrac{b^2 + c^2 - a^2}{2bc}$ (if a, b, c are small)

      2. $\tan \dfrac{A}{2} = \sqrt{\dfrac{(s-b)(s-c)}{s(s-a)}}$ (for use with logs)

         $\sin \dfrac{A}{2} = \sqrt{\dfrac{(s-b)(s-c)}{bc}}$

         $\cos \dfrac{A}{2} = \sqrt{\dfrac{s(s-a)}{bc}}$

         $\sin A = \dfrac{2}{bc} \sqrt{s(s-a)(s-b)(s-c)}.$

  Case II.   Two sides and contained angle known.

$$\tan \frac{B-C}{2} = \frac{b-c}{b+c} \cot \frac{A}{2}.$$

  Case III.   Two angles and a side known.

$$\frac{\sin A}{a} = \frac{\sin B}{b} = \frac{\sin C}{c}$$

7. Ratios of angles between 0 and 2π radians.

$\sin \theta = \sin(\pi - \theta) = -\sin(\pi + \theta) = -\sin(2\pi - \theta)$
$\cos\ \ \ = -\cos(\pi - \theta) = -\cos(\pi + \theta) = \cos(2\pi - \theta)$
$\tan\ \ \ = -\tan(\pi - \theta) = \tan(\pi + \theta) = -\tan(2\pi - \theta)$

8. Ratios of θ and $-\theta$.

$$\sin \theta = -\sin(-\theta)$$
$$\cos \theta = \cos(-\theta)$$
$$\tan \theta = -\tan(-\theta)$$

9. General formulae for angles with the same ratios as θ.

     sine $n\pi + (-1)^n\theta$
     cosine $2n\pi \pm \theta$
     tangent $n\pi + \theta.$

10. Circular measure.

      1 radian $= 57°\ 17'\ 45''$ (approx.)

  To convert degrees to radians.

$$\theta° = \left(\theta° \times \frac{\pi}{180}\right) \text{ radians.}$$

  Length of an arc.

$$a = r\theta \ (\theta \text{ in radians}).$$

LOGARITHMS of numbers 100 to 549

Proportional Parts

	0	1	2	3	4	5	6	7	8	9	1	2	3	4	5	6	7	8	9
10	0000	0043	0086	0128	0170	0212	0253	0294	0334	0374	4	8	12	17	21	25	29	33	37
11	0414	0453	0492	0531	0569	0607	0645	0682	0719	0755	4	8	11	15	19	23	26	30	34
12	0792	0828	0864	0899	0934	0969	1004	1038	1072	1106	3	7	10	14	17	21	24	28	31
13	1139	1173	1206	1239	1271	1303	1335	1367	1399	1430	3	6	10	13	16	19	23	26	29
14	1461	1492	1523	1553	1584	1614	1644	1673	1703	1732	3	6	9	12	15	18	21	24	27
15	1761	1790	1818	1847	1875	1903	1931	1959	1987	2014	3	6	8	11	14	17	20	22	25
16	2041	2068	2095	2122	2148	2175	2201	2227	2253	2279	3	5	8	11	13	16	18	21	24
17	2304	2330	2355	2380	2405	2430	2455	2480	2504	2529	2	5	7	10	12	15	17	20	22
18	2553	2577	2601	2625	2648	2672	2695	2718	2742	2765	2	5	7	9	12	14	16	19	21
19	2788	2810	2833	2856	2878	2900	2923	2945	2967	2989	2	4	7	9	11	13	16	18	20
20	3010	3032	3054	3075	3096	3118	3139	3160	3181	3201	2	4	6	8	11	13	15	17	19
21	3222	3243	3263	3284	3304	3324	3345	3365	3385	3404	2	4	6	8	10	12	14	16	18
22	3424	3444	3464	3483	3502	3522	3541	3560	3579	3598	2	4	6	8	10	12	14	15	17
23	3617	3636	3655	3674	3692	3711	3729	3747	3766	3784	2	4	6	7	9	11	13	15	17
24	3802	3820	3838	3856	3874	3892	3909	3927	3945	3962	2	4	5	7	9	11	12	14	16
25	3979	3997	4014	4031	4048	4065	4082	4099	4116	4133	2	3	5	7	9	10	12	14	15
26	4150	4166	4183	4200	4216	4232	4249	4265	4281	4298	2	3	5	7	8	10	11	13	15
27	4314	4330	4346	4362	4378	4393	4409	4425	4440	4456	2	3	5	6	8	9	11	13	14
28	4472	4487	4502	4518	4533	4548	4564	4579	4594	4609	2	3	5	6	8	9	11	12	14
29	4624	4639	4654	4669	4683	4698	4713	4728	4742	4757	1	3	4	6	7	9	10	12	13
30	4771	4786	4800	4814	4829	4843	4857	4871	4886	4900	1	3	4	6	7	9	10	11	13
31	4914	4928	4942	4955	4969	4983	4997	5011	5024	5038	1	3	4	5	7	8	10	11	12
32	5051	5065	5079	5092	5105	5119	5132	5145	5159	5172	1	3	4	5	7	8	9	11	12
33	5185	5198	5211	5224	5237	5250	5263	5276	5289	5302	1	3	4	5	6	8	9	10	12
34	5315	5328	5340	5353	5366	5378	5391	5403	5416	5428	1	3	4	5	6	8	9	10	11
35	5441	5453	5465	5478	5490	5502	5514	5527	5539	5551	1	2	4	5	6	7	9	10	11
36	5563	5575	5587	5599	5611	5623	5635	5647	5658	5670	1	2	4	5	6	7	8	10	11
37	5682	5694	5705	5717	5729	5740	5752	5763	5775	5786	1	2	3	5	6	7	8	9	10
38	5798	5809	5821	5832	5843	5855	5866	5877	5888	5899	1	2	3	5	6	7	8	9	10
39	5911	5922	5933	5944	5955	5966	5977	5988	5999	6010	1	2	3	4	5	7	8	9	10
40	6021	6031	6042	6053	6064	6075	6085	6096	6107	6117	1	2	3	4	5	6	7	9	10
41	6128	6138	6149	6160	6170	6180	6191	6201	6212	6222	1	2	3	4	5	6	7	8	9
42	6232	6243	6253	6263	6274	6284	6294	6304	6314	6325	1	2	3	4	5	6	7	8	9
43	6335	6345	6355	6365	6375	6385	6395	6405	6415	6425	1	2	3	4	5	6	7	8	9
44	6435	6444	6454	6464	6474	6484	6493	6503	6513	6522	1	2	3	4	5	6	7	8	9
45	6532	6542	6551	6561	6571	6580	6590	6599	6609	6618	1	2	3	4	5	6	7	8	9
46	6628	6637	6646	6656	6665	6675	6684	6693	6702	6712	1	2	3	4	5	6	7	7	8
47	6721	6730	6739	6749	6758	6767	6776	6785	6794	6803	1	2	3	4	5	5	6	7	8
48	6812	6821	6830	6839	6848	6857	6866	6875	6884	6893	1	2	3	4	4	5	6	7	8
49	6902	6911	6920	6928	6937	6946	6955	6964	6972	6981	1	2	3	4	4	5	6	7	8
50	6990	6998	7007	7016	7024	7033	7042	7050	7059	7067	1	2	3	3	4	5	6	7	8
51	7076	7084	7093	7101	7110	7118	7126	7135	7143	7152	1	2	3	3	4	5	6	7	8
52	7160	7168	7177	7185	7193	7202	7210	7218	7226	7235	1	2	2	3	4	5	6	7	7
53	7243	7251	7259	7267	7275	7284	7292	7300	7308	7316	1	2	2	3	4	5	6	6	7
54	7324	7332	7340	7348	7356	7364	7372	7380	7388	7396	1	2	2	3	4	5	6	6	7
	0	1	2	3	4	5	6	7	8	9	1	2	3	4	5	6	7	8	9

LOGARITHMS of numbers 550 to 999

	0	1	2	3	4	5	6	7	8	9	1	2	3	4	5	6	7	8	9
55	7404	7412	7419	7427	7435	7443	7451	7459	7466	7474	1	2	2	3	4	5	5	6	7
56	7482	7490	7497	7505	7513	7520	7528	7536	7543	7551	1	2	2	3	4	5	5	6	7
57	7559	7566	7574	7582	7589	7597	7604	7612	7619	7627	1	2	2	3	4	5	5	6	7
58	7634	7642	7649	7657	7664	7672	7679	7686	7694	7701	1	1	2	3	4	4	5	6	7
59	7709	7716	7723	7731	7738	7745	7752	7760	7767	7774	1	1	2	3	4	4	5	6	7
60	7782	7789	7796	7803	7810	7818	7825	7832	7839	7846	1	1	2	3	4	4	5	6	6
61	7853	7860	7868	7875	7882	7889	7896	7903	7910	7917	1	1	2	3	4	4	5	6	6
62	7924	7931	7938	7945	7952	7959	7966	7973	7980	7987	1	1	2	3	3	4	5	6	6
63	7993	8000	8007	8014	8021	8028	8035	8041	8048	8055	1	1	2	3	3	4	5	6	6
64	8062	8069	8075	8082	8089	8096	8102	8109	8116	8122	1	1	2	3	3	4	5	5	6
65	8129	8136	8142	8149	8156	8162	8169	8176	8182	8189	1	1	2	3	3	4	5	5	6
66	8195	8202	8209	8215	8222	8228	8235	8241	8248	8254	1	1	2	3	3	4	5	5	6
67	8261	8267	8274	8280	8287	8293	8299	8306	8312	8319	1	1	2	3	3	4	4	5	6
68	8325	8331	8338	8344	8351	8357	8363	8370	8376	8382	1	1	2	3	3	4	4	5	6
69	8388	8395	8401	8407	8414	8420	8426	8432	8439	8445	1	1	2	3	3	4	4	5	6
70	8451	8457	8463	8470	8476	8482	8488	8494	8500	8506	1	1	2	2	3	4	4	5	6
71	8513	8519	8525	8531	8537	8543	8549	8555	8561	8567	1	1	2	2	3	4	4	5	5
72	8573	8579	8585	8591	8597	8603	8609	8615	8621	8627	1	1	2	2	3	4	4	5	5
73	8633	8639	8645	8651	8657	8663	8669	8675	8681	8686	1	1	2	2	3	4	4	5	5
74	8692	8698	8704	8710	8716	8722	8727	8733	8739	8745	1	1	2	2	3	4	4	5	5
75	8751	8756	8762	8768	8774	8779	8785	8791	8797	8802	1	1	2	2	3	3	4	5	5
76	8808	8814	8820	8825	8831	8837	8842	8848	8854	8859	1	1	2	2	3	3	4	5	5
77	8865	8871	8876	8882	8887	8893	8899	8904	8910	8915	1	1	2	2	3	3	4	4	5
78	8921	8927	8932	8938	8943	8949	8954	8960	8965	8971	1	1	2	2	3	3	4	4	5
79	8976	8982	8987	8993	8998	9004	9009	9015	9020	9025	1	1	2	2	3	3	4	4	5
80	9031	9036	9042	9047	9053	9058	9063	9069	9074	9079	1	1	2	2	3	3	4	4	5
81	9085	9090	9096	9101	9106	9112	9117	9122	9128	9133	1	1	2	2	3	3	4	4	5
82	9138	9143	9149	9154	9159	9165	9170	9175	9180	9186	1	1	2	2	3	3	4	4	5
83	9191	9196	9201	9206	9212	9217	9222	9227	9232	9238	1	1	2	2	3	3	4	4	5
84	9243	9248	9253	9258	9263	9269	9274	9279	9284	9289	1	1	2	2	3	3	4	4	5
85	9294	9299	9304	9309	9315	9320	9325	9330	9335	9340	1	1	2	2	3	3	4	4	5
86	9345	9350	9355	9360	9365	9370	9375	9380	9385	9390	1	1	2	2	3	3	4	4	5
87	9395	9400	9405	9410	9415	9420	9425	9430	9435	9440	0	1	1	2	2	3	3	4	4
88	9445	9450	9455	9460	9465	9469	9474	9479	9484	9489	0	1	1	2	2	3	3	4	4
89	9494	9499	9504	9509	9513	9518	9523	9528	9533	9538	0	1	1	2	2	3	3	4	4
90	9542	9547	9552	9557	9562	9566	9571	9576	9581	9586	0	1	1	2	2	3	3	4	4
91	9590	9595	9600	9605	9609	9614	9619	9624	9628	9633	0	1	1	2	2	3	3	4	4
92	9638	9643	9647	9652	9657	9661	9666	9671	9675	9680	0	1	1	2	2	3	3	4	4
93	9685	9689	9694	9699	9703	9708	9713	9717	9722	9727	0	1	1	2	2	3	3	4	4
94	9731	9736	9741	9745	9750	9754	9759	9764	9768	9773	0	1	1	2	2	3	3	4	4
95	9777	9782	9786	9791	9795	9800	9805	9809	9814	9818	0	1	1	2	2	3	3	4	4
96	9823	9827	9832	9836	9841	9845	9850	9854	9859	9863	0	1	1	2	2	3	3	4	4
97	9868	9872	9877	9881	9886	9890	9894	9899	9903	9908	0	1	1	2	2	3	3	4	4
98	9912	9917	9921	9926	9930	9934	9939	9943	9948	9952	0	1	1	2	2	3	3	4	4
99	9956	9961	9965	9969	9974	9978	9983	9987	9991	9996	0	1	1	2	2	3	3	4	4
	0	1	2	3	4	5	6	7	8	9	1	2	3	4	5	6	7	8	9

ANTI-LOGARITHMS

	0	1	2	3	4	5	6	7	8	9	1	2	3	4	5	6	7	8	9
·00	1000	1002	1005	1007	1009	1012	1014	1016	1019	1021	0	0	1	1	1	1	2	2	2
·01	1023	1026	1028	1030	1033	1035	1038	1040	1042	1045	0	0	1	1	1	1	2	2	2
·02	1047	1050	1052	1054	1057	1059	1062	1064	1067	1069	0	0	1	1	1	1	2	2	2
·03	1072	1074	1076	1079	1081	1084	1086	1089	1091	1094	0	0	1	1	1	1	2	2	2
·04	1096	1099	1102	1104	1107	1109	1112	1114	1117	1119	0	1	1	1	1	2	2	2	2
·05	1122	1125	1127	1130	1132	1135	1138	1140	1143	1146	0	1	1	1	1	2	2	2	2
·06	1148	1151	1153	1156	1159	1161	1164	1167	1169	1172	0	1	1	1	1	2	2	2	2
·07	1175	1178	1180	1183	1186	1189	1191	1194	1197	1199	0	1	1	1	1	2	2	2	2
·08	1202	1205	1208	1211	1213	1216	1219	1222	1225	1227	0	1	1	1	1	2	2	2	3
·09	1230	1233	1236	1239	1242	1245	1247	1250	1253	1256	0	1	1	1	1	2	2	2	3
·10	1259	1262	1265	1268	1271	1274	1276	1279	1282	1285	0	1	1	1	1	2	2	2	3
·11	1288	1291	1294	1297	1300	1303	1306	1309	1312	1315	0	1	1	1	2	2	2	2	3
·12	1318	1321	1324	1327	1330	1334	1337	1340	1343	1346	0	1	1	1	2	2	2	3	3
·13	1349	1352	1355	1358	1361	1365	1368	1371	1374	1377	0	1	1	1	2	2	2	3	3
·14	1380	1384	1387	1390	1393	1396	1400	1403	1406	1409	0	1	1	1	2	2	2	3	3
·15	1413	1416	1419	1422	1426	1429	1432	1435	1439	1442	0	1	1	1	2	2	2	3	3
·16	1445	1449	1452	1455	1459	1462	1466	1469	1472	1476	0	1	1	1	2	2	2	3	3
·17	1479	1483	1486	1489	1493	1496	1500	1503	1507	1510	0	1	1	1	2	2	2	3	3
·18	1514	1517	1521	1524	1528	1531	1535	1538	1542	1545	0	1	1	1	2	2	2	3	3
·19	1549	1552	1556	1560	1563	1567	1570	1574	1578	1581	0	1	1	1	2	2	3	3	3
·20	1585	1589	1592	1596	1600	1603	1607	1611	1614	1618	0	1	1	1	2	2	3	3	3
·21	1622	1626	1629	1633	1637	1641	1644	1648	1652	1656	0	1	1	2	2	2	3	3	3
·22	1660	1663	1667	1671	1675	1679	1683	1687	1690	1694	0	1	1	2	2	2	3	3	3
·23	1698	1702	1706	1710	1714	1718	1722	1726	1730	1734	0	1	1	2	2	2	3	3	4
·24	1738	1742	1746	1750	1754	1758	1762	1766	1770	1774	0	1	1	2	2	2	3	3	4
·25	1778	1782	1786	1791	1795	1799	1803	1807	1811	1816	0	1	1	2	2	3	3	3	4
·26	1820	1824	1828	1832	1837	1841	1845	1849	1854	1858	0	1	1	2	2	3	3	3	4
·27	1862	1866	1871	1875	1879	1884	1888	1892	1897	1901	0	1	1	2	2	3	3	3	4
·28	1905	1910	1914	1919	1923	1928	1932	1936	1941	1945	0	1	1	2	2	3	3	4	4
·29	1950	1954	1959	1963	1968	1972	1977	1982	1986	1991	0	1	1	2	2	3	3	4	4
·30	1995	2000	2004	2009	2014	2018	2023	2028	2032	2037	0	1	1	2	2	3	3	4	4
·31	2042	2046	2051	2056	2061	2065	2070	2075	2080	2084	0	1	1	2	2	3	3	4	4
·32	2089	2094	2099	2104	2109	2113	2118	2123	2128	2133	0	1	1	2	2	3	3	4	4
·33	2138	2143	2148	2153	2158	2163	2168	2173	2178	2183	0	1	1	2	2	3	3	4	4
·34	2188	2193	2198	2203	2208	2213	2218	2223	2228	2234	1	1	2	2	3	3	4	4	5
·35	2239	2244	2249	2254	2259	2265	2270	2275	2280	2286	1	1	2	2	3	3	4	4	5
·36	2291	2296	2301	2307	2312	2317	2323	2328	2333	2339	1	1	2	2	3	3	4	4	5
·37	2344	2350	2355	2360	2366	2371	2377	2382	2388	2393	1	1	2	2	3	3	4	4	5
·38	2399	2404	2410	2415	2421	2427	2432	2438	2443	2449	1	1	2	2	3	3	4	4	5
·39	2455	2460	2466	2472	2477	2483	2489	2495	2500	2506	1	1	2	2	3	3	4	5	5
·40	2512	2518	2523	2529	2535	2541	2547	2553	2559	2564	1	1	2	2	3	3	4	5	5
·41	2570	2576	2582	2588	2594	2600	2606	2612	2618	2624	1	1	2	2	3	4	4	5	5
·42	2630	2636	2642	2648	2655	2661	2667	2673	2679	2685	1	1	2	2	3	4	4	5	6
·43	2692	2698	2704	2710	2716	2723	2729	2735	2742	2748	1	1	2	2	3	4	4	5	6
·44	2754	2761	2767	2773	2780	2786	2793	2799	2805	2812	1	1	2	3	3	4	4	5	6
·45	2818	2825	2831	2838	2844	2851	2858	2864	2871	2877	1	1	2	3	3	4	5	5	6
·46	2884	2891	2897	2904	2911	2917	2924	2931	2938	2944	1	1	2	3	3	4	5	5	6
·47	2951	2958	2965	2972	2979	2985	2992	2999	3006	3013	1	1	2	3	3	4	5	6	6
·48	3020	3027	3034	3041	3048	3055	3062	3069	3076	3083	1	1	2	3	4	4	5	6	6
·49	3090	3097	3105	3112	3119	3126	3133	3141	3148	3155	1	1	2	3	4	4	5	6	7
	0	1	2	3	4	5	6	7	8	9	1	2	3	4	5	6	7	8	9

ANTI-LOGARITHMS

Proportional Parts

	0	1	2	3	4	5	6	7	8	9	1	2	3	4	5	6	7	8	9
·50	3162	3170	3177	3184	3192	3199	3206	3214	3221	3228	1	1	2	3	4	4	5	6	7
·51	3236	3243	3251	3258	3266	3273	3281	3289	3296	3304	1	2	2	3	4	5	5	6	7
·52	3311	3319	3327	3334	3342	3350	3357	3365	3373	3381	1	2	2	3	4	5	5	6	7
·53	3388	3396	3404	3412	3420	3428	3436	3443	3451	3459	1	2	2	3	4	5	6	6	7
·54	3467	3475	3483	3491	3499	3508	3516	3524	3532	3540	1	2	2	3	4	5	6	6	7
·55	3548	3556	3565	3573	3581	3589	3597	3606	3614	3622	1	2	2	3	4	5	6	7	7
·56	3631	3639	3648	3656	3664	3673	3681	3690	3698	3707	1	2	3	3	4	5	6	7	8
·57	3715	3724	3733	3741	3750	3758	3767	3776	3784	3793	1	2	3	3	4	5	6	7	8
·58	3802	3811	3819	3828	3837	3846	3855	3864	3873	3882	1	2	3	4	4	5	6	7	8
·59	3890	3899	3908	3917	3926	3936	3945	3954	3963	3972	1	2	3	4	5	5	6	7	8
·60	3981	3990	3999	4009	4018	4027	4036	4046	4055	4064	1	2	3	4	5	6	7	7	8
·61	4074	4083	4093	4102	4111	4121	4130	4140	4150	4159	1	2	3	4	5	6	7	8	9
·62	4169	4178	4188	4198	4207	4217	4227	4236	4246	4256	1	2	3	4	5	6	7	8	9
·63	4266	4276	4285	4295	4305	4315	4325	4335	4345	4355	1	2	3	4	5	6	7	8	9
·64	4365	4375	4385	4395	4406	4416	4426	4436	4446	4457	1	2	3	4	5	6	7	8	9
·65	4467	4477	4487	4498	4508	4519	4529	4539	4550	4560	1	2	3	4	5	6	7	8	9
·66	4571	4581	4592	4603	4613	4624	4634	4645	4656	4667	1	2	3	4	5	6	7	8	10
·67	4677	4688	4699	4710	4721	4732	4742	4753	4764	4775	1	2	3	4	5	7	8	9	10
·68	4786	4797	4808	4819	4831	4842	4853	4864	4875	4887	1	2	3	4	6	7	8	9	10
·69	4898	4909	4920	4932	4943	4955	4966	4977	4989	5000	1	2	3	5	6	7	8	9	10
·70	5012	5023	5035	5047	5058	5070	5082	5093	5105	5117	1	2	4	5	6	7	8	9	11
·71	5129	5140	5152	5164	5176	5188	5200	5212	5224	5236	1	2	4	5	6	7	8	10	11
·72	5248	5260	5272	5284	5297	5309	5321	5333	5346	5358	1	2	4	5	6	7	9	10	11
·73	5370	5383	5395	5408	5420	5433	5445	5458	5470	5483	1	3	4	5	6	8	9	10	11
·74	5495	5508	5521	5534	5546	5559	5572	5585	5598	5610	1	3	4	5	6	8	9	10	12
·75	5623	5636	5649	5662	5675	5689	5702	5715	5728	5741	1	3	4	5	7	8	9	10	12
·76	5754	5768	5781	5794	5808	5821	5834	5848	5861	5875	1	3	4	5	7	8	9	11	12
·77	5888	5902	5916	5929	5943	5957	5970	5984	5998	6012	1	3	4	6	7	8	10	11	12
·78	6026	6039	6053	6067	6081	6095	6109	6124	6138	6152	1	3	4	6	7	8	10	11	13
·79	6166	6180	6194	6209	6223	6237	6252	6266	6281	6295	1	3	4	6	7	9	10	12	13
·80	6310	6324	6339	6353	6368	6383	6397	6412	6427	6442	1	3	4	6	7	9	10	12	13
·81	6457	6471	6486	6501	6516	6531	6546	6561	6577	6592	2	3	5	6	8	9	11	12	14
·82	6607	6622	6637	6653	6668	6683	6699	6714	6730	6745	2	3	5	6	8	9	11	12	14
·83	6761	6776	6792	6808	6823	6839	6855	6871	6887	6902	2	3	5	6	8	9	11	13	14
·84	6918	6934	6950	6966	6982	6998	7015	7031	7047	7063	2	3	5	6	8	10	11	13	14
·85	7079	7096	7112	7129	7145	7161	7178	7194	7211	7228	2	3	5	7	8	10	12	13	15
·86	7244	7261	7278	7295	7311	7328	7345	7362	7379	7396	2	3	5	7	8	10	12	14	15
·87	7413	7430	7447	7464	7482	7499	7516	7534	7551	7568	2	3	5	7	9	10	12	14	16
·88	7586	7603	7621	7638	7656	7674	7691	7709	7727	7745	2	4	5	7	9	11	12	14	16
·89	7762	7780	7798	7816	7834	7852	7870	7889	7907	7925	2	4	5	7	9	11	13	14	16
·90	7943	7962	7980	7998	8017	8035	8054	8072	8091	8110	2	4	6	7	9	11	13	15	17
·91	8128	8147	8166	8185	8204	8222	8241	8260	8279	8299	2	4	6	8	10	11	13	15	17
·92	8318	8337	8356	8375	8395	8414	8433	8453	8472	8492	2	4	6	8	10	12	14	15	17
·93	8511	8531	8551	8570	8590	8610	8630	8650	8670	8690	2	4	6	8	10	12	14	16	18
·94	8710	8730	8750	8770	8790	8810	8831	8851	8872	8892	2	4	6	8	10	12	14	16	18
·95	8913	8933	8954	8974	8995	9016	9036	9057	9078	9099	2	4	6	8	10	12	14	17	19
·96	9120	9141	9162	9183	9204	9226	9247	9268	9290	9311	2	4	6	9	11	13	15	17	19
·97	9333	9354	9376	9397	9419	9441	9462	9484	9506	9528	2	4	7	9	11	13	15	17	20
·98	9550	9572	9594	9616	9638	9661	9683	9705	9727	9750	2	4	7	9	11	13	16	18	20
·99	9772	9795	9817	9840	9863	9886	9908	9931	9954	9977	2	5	7	9	11	14	16	18	21
	0	1	2	3	4	5	6	7	8	9	1	2	3	4	5	6	7	8	9

NATURAL SINES

	0'	6'	12'	18'	24'	30'	36'	42'	48'	54'	1'	2'	3'	4'	5'
0°	0·0000	·0017	·0035	·0052	·0070	·0087	·0105	·0122	·0140	·0157	3	6	9	12	15
1	0·0175	·0192	·0209	·0227	·0244	·0262	·0279	·0297	·0314	·0332	3	6	9	12	15
2	0·0349	·0366	·0384	·0401	·0419	·0436	·0454	·0471	·0489	·0506	3	6	9	12	15
3	0·0523	·0541	·0558	·0576	·0593	·0610	·0628	·0645	·0663	·0680	3	6	9	12	15
4	0·0698	·0715	·0732	·0750	·0767	·0785	·0802	·0819	·0837	·0854	3	6	9	12	14
5	0·0872	·0889	·0906	·0924	·0941	·0958	·0976	·0993	·1011	·1028	3	6	9	12	14
6	0·1045	·1063	·1080	·1097	·1115	·1132	·1149	·1167	·1184	·1201	3	6	9	12	14
7	0·1219	·1236	·1253	·1271	·1288	·1305	·1323	·1340	·1357	·1374	3	6	9	12	14
8	0·1392	·1409	·1426	·1444	·1461	·1478	·1495	·1513	·1530	·1547	3	6	9	11	14
9	0·1564	·1582	·1599	·1616	·1633	·1650	·1668	·1685	·1702	·1719	3	6	9	11	14
10	0·1736	·1754	·1771	·1788	·1805	·1822	·1840	·1857	·1874	·1891	3	6	9	11	14
11	0·1908	·1925	·1942	·1959	·1977	·1994	·2011	·2028	·2045	·2062	3	6	9	11	14
12	0·2079	·2096	·2113	·2130	·2147	·2164	·2181	·2198	·2215	·2232	3	6	9	11	14
13	0·2250	·2267	·2284	·2300	·2317	·2334	·2351	·2368	·2385	·2402	3	6	8	11	14
14	0·2419	·2436	·2453	·2470	·2487	·2504	·2521	·2538	·2554	·2571	3	6	8	11	14
15	0·2588	·2605	·2622	·2639	·2656	·2672	·2689	·2706	·2723	·2740	3	6	8	11	14
16	0·2756	·2773	·2790	·2807	·2823	·2840	·2857	·2874	·2890	·2907	3	6	8	11	14
17	0·2924	·2940	·2957	·2974	·2990	·3007	·3024	·3040	·3057	·3074	3	6	8	11	14
18	0·3090	·3107	·3123	·3140	·3156	·3173	·3190	·3206	·3223	·3239	3	6	8	11	14
19	0·3256	·3272	·3289	·3305	·3322	·3338	·3355	·3371	·3387	·3404	3	5	8	11	14
20	0·3420	·3437	·3453	·3469	·3486	·3502	·3518	·3535	·3551	·3567	3	5	8	11	14
21	0·3584	·3600	·3616	·3633	·3649	·3665	·3681	·3697	·3714	·3730	3	5	8	11	14
22	0·3746	·3762	·3778	·3795	·3811	·3827	·3843	·3859	·3875	·3891	3	5	8	11	13
23	0·3907	·3923	·3939	·3955	·3971	·3987	·4003	·4019	·4035	·4051	3	5	8	11	13
24	0·4067	·4083	·4099	·4115	·4131	·4147	·4163	·4179	·4195	·4210	3	5	8	11	13
25	0·4226	·4242	·4258	·4274	·4289	·4305	·4321	·4337	·4352	·4368	3	5	8	11	13
26	0·4384	·4399	·4415	·4431	·4446	·4462	·4478	·4493	·4509	·4524	3	5	8	10	13
27	0·4540	·4555	·4571	·4586	·4602	·4617	·4633	·4648	·4664	·4679	3	5	8	10	13
28	0·4695	·4710	·4726	·4741	·4756	·4772	·4787	·4802	·4818	·4833	3	5	8	10	13
29	0·4848	·4863	·4879	·4894	·4909	·4924	·4939	·4955	·4970	·4985	3	5	8	10	13
30	0·5000	·5015	·5030	·5045	·5060	·5075	·5090	·5105	·5120	·5135	2	5	8	10	12
31	0·5150	·5165	·5180	·5195	·5210	·5225	·5240	·5255	·5270	·5284	2	5	7	10	12
32	0·5299	·5314	·5329	·5344	·5358	·5373	·5388	·5402	·5417	·5432	2	5	7	10	12
33	0·5446	·5461	·5476	·5490	·5505	·5519	·5534	·5548	·5563	·5577	2	5	7	10	12
34	0·5592	·5606	·5621	·5635	·5650	·5664	·5678	·5693	·5707	·5721	2	5	7	10	12
35	0·5736	·5750	·5764	·5779	·5793	·5807	·5821	·5835	·5850	·5864	2	5	7	9	12
36	0·5878	·5892	·5906	·5920	·5934	·5948	·5962	·5976	·5990	·6004	2	5	7	9	12
37	0·6018	·6032	·6046	·6060	·6074	·6088	·6101	·6115	·6129	·6143	2	5	7	9	12
38	0·6157	·6170	·6184	·6198	·6211	·6225	·6239	·6252	·6266	·6280	2	5	7	9	11
39	0·6293	·6307	·6320	·6334	·6347	·6361	·6374	·6388	·6401	·6414	2	4	7	9	11
40	0·6428	·6441	·6455	·6468	·6481	·6494	·6508	·6521	·6534	·6547	2	4	7	9	11
41	0·6561	·6574	·6587	·6600	·6613	·6626	·6639	·6652	·6665	·6678	2	4	6	9	11
42	0·6691	·6704	·6717	·6730	·6743	·6756	·6769	·6782	·6794	·6807	2	4	6	9	11
43	0·6820	·6833	·6845	·6858	·6871	·6884	·6896	·6909	·6921	·6934	2	4	6	8	11
44	0·6947	·6959	·6972	·6984	·6997	·7009	·7022	·7034	·7046	·7059	2	4	6	8	10
	0'	6'	12'	18'	24'	30'	36'	42'	48'	54'	1'	2'	3'	4'	5'

NATURAL SINES

	0′	6′	12′	18′	24′	30′	36′	42′	48′	54′	1′	2′	3′	4′	5′
45°	0·7071	·7083	·7096	·7108	·7120	·7133	·7145	·7157	·7169	·7181	2	4	6	8	10
46	0·7193	·7206	·7218	·7230	·7242	·7254	·7266	·7278	·7290	·7302	2	4	6	8	10
47	0·7314	·7325	·7337	·7349	·7361	·7373	·7385	·7396	·7408	·7420	2	4	6	8	10
48	0·7431	·7443	·7455	·7466	·7478	·7490	·7501	·7513	·7524	·7536	2	4	6	8	10
49	0·7547	·7559	·7570	·7581	·7593	·7604	·7615	·7627	·7638	·7649	2	4	6	8	9
50	0·7660	·7672	·7683	·7694	·7705	·7716	·7727	·7738	·7749	·7760	2	4	6	7	9
51	0·7771	·7782	·7793	·7804	·7815	·7826	·7837	·7848	·7859	·7869	2	4	5	7	9
52	0·7880	·7891	·7902	·7912	·7923	·7934	·7944	·7955	·7965	·7976	2	4	5	7	9
53	0·7986	·7997	·8007	·8018	·8028	·8039	·8049	·8059	·8070	·8080	2	3	5	7	9
54	0·8090	·8100	·8111	·8121	·8131	·8141	·8151	·8161	·8171	·8181	2	3	5	7	8
55	0·8192	·8202	·8211	·8221	·8231	·8241	·8251	·8261	·8271	·8281	2	3	5	7	8
56	0·8290	·8300	·8310	·8320	·8329	·8339	·8348	·8358	·8368	·8377	2	3	5	6	8
57	0·8387	·8396	·8406	·8415	·8425	·8434	·8443	·8453	·8462	·8471	2	3	5	6	8
58	0·8480	·8490	·8499	·8508	·8517	·8526	·8536	·8545	·8554	·8563	2	3	5	6	8
59	0·8572	·8581	·8590	·8599	·8607	·8616	·8625	·8634	·8643	·8652	1	3	4	6	7
60	0·8660	·8669	·8678	·8686	·8695	·8704	·8712	·8721	·8729	·8738	1	3	4	6	7
61	0·8746	·8755	·8763	·8771	·8780	·8788	·8796	·8805	·8813	·8821	1	3	4	6	7
62	0·8829	·8838	·8846	·8854	·8862	·8870	·8878	·8886	·8894	·8902	1	3	4	5	7
63	0·8910	·8918	·8926	·8934	·8942	·8949	·8957	·8965	·8973	·8980	1	3	4	5	6
64	0·8988	·8996	·9003	·9011	·9018	·9026	·9033	·9041	·9048	·9056	1	2	4	5	6
65	0·9063	·9070	·9078	·9085	·9092	·9100	·9107	·9114	·9121	·9128	1	2	4	5	6
66	0·9135	·9143	·9150	·9157	·9164	·9171	·9178	·9184	·9191	·9198	1	2	3	5	6
67	0·9205	·9212	·9219	·9225	·9232	·9239	·9245	·9252	·9259	·9265	1	2	3	4	6
68	0·9272	·9278	·9285	·9291	·9298	·9304	·9311	·9317	·9323	·9330	1	2	3	4	5
69	0·9336	·9342	·9348	·9354	·9361	·9367	·9373	·9379	·9385	·9391	1	2	3	4	5
70	0·9397	·9403	·9409	·9415	·9421	·9426	·9432	·9438	·9444	·9449	1	2	3	4	5
71	0·9455	·9461	·9466	·9472	·9478	·9483	·9489	·9494	·9500	·9505	1	2	3	4	5
72	0·9511	·9516	·9521	·9527	·9532	·9537	·9542	·9548	·9553	·9558	1	2	3	3	4
73	0·9563	·9568	·9573	·9578	·9583	·9588	·9593	·9598	·9603	·9608	1	2	2	3	4
74	0·9613	·9617	·9622	·9627	·9632	·9636	·9641	·9646	·9650	·9655	1	2	2	3	4
75	0·9659	·9664	·9668	·9673	·9677	·9681	·9686	·9690	·9694	·9699	1	1	2	3	4
76	0·9703	·9707	·9711	·9715	·9720	·9724	·9728	·9732	·9736	·9740	1	1	2	3	3
77	0·9744	·9748	·9751	·9755	·9759	·9763	·9767	·9770	·9774	·9778	1	1	2	2	3
78	0·9781	·9785	·9789	·9792	·9796	·9799	·9803	·9806	·9810	·9813	1	1	2	2	3
79	0·9816	·9820	·9823	·9826	·9829	·9833	·9836	·9839	·9842	·9845	1	1	2	2	3
80	0·9848	·9851	·9854	·9857	·9860	·9863	·9866	·9869	·9871	·9874	0	1	1	2	2
81	0·9877	·9880	·9882	·9885	·9888	·9890	·9893	·9895	·9898	·9900	0	1	1	2	2
82	0·9903	·9905	·9907	·9910	·9912	·9914	·9917	·9919	·9921	·9923	0	1	1	1	2
83	0·9925	·9928	·9930	·9932	·9934	·9936	·9938	·9940	·9942	·9943	0	1	1	1	2
84	0·9945	·9947	·9949	·9951	·9952	·9954	·9956	·9957	·9959	·9960	0	1	1	1	1
85	0·9962	·9963	·9965	·9966	·9968	·9969	·9971	·9972	·9973	·9974	0	0	1	1	1
86	0·9976	·9977	·9978	·9979	·9980	·9981	·9982	·9983	·9984	·9985	0	0	0	1	1
87	0·9986	·9987	·9988	·9989	·9990	·9990	·9991	·9992	·9993	·9993	0	0	0	1	1
88	0·9994	·9995	·9995	·9996	·9996	·9997	·9997	·9997	·9998	·9998	0	0	0	0	0
89	0·9998	·9999	·9999	·9999	0·9999	1·0000	·0000	·0000	·0000	·0000	0	0	0	0	0
	0′	6′	12′	18′	24′	30′	36′	42′	48′	54′	1′	2′	3′	4′	5′

NATURAL COSINES

	0′	6′	12′	18′	24′	30′	36′	42′	48′	54′	1′	2′	3′	4′	5′
0°	1·0000	·0000	·0000	·0000	·0000	1·0000	0·9999	·9999	·9999	·9999	0	0	0	0	0
1	0·9998	·9998	·9998	·9997	·9997	·9997	·9996	·9996	·9995	·9995	0	0	0	0	0
2	0·9994	·9993	·9993	·9992	·9991	·9990	·9990	·9989	·9988	·9987	0	0	0	0	1
3	0·9986	·9985	·9984	·9983	·9982	·9981	·9980	·9979	·9978	·9977	0	0	0	1	1
4	0·9976	·9974	·9973	·9972	·9971	·9969	·9968	·9966	·9965	·9963	0	0	1	1	1
5	0·9962	·9960	·9959	·9957	·9956	·9954	·9952	·9951	·9949	·9947	0	1	1	1	1
6	0·9945	·9943	·9942	·9940	·9938	·9936	·9934	·9932	·9930	·9928	0	1	1	1	2
7	0·9925	·9923	·9921	·9919	·9917	·9914	·9912	·9910	·9907	·9905	0	1	1	1	2
8	0·9903	·9900	·9898	·9895	·9893	·9890	·9888	·9885	·9882	·9880	0	1	1	2	2
9	0·9877	·9874	·9871	·9869	·9866	·9863	·9860	·9857	·9854	·9851	0	1	1	2	2
10	0·9848	·9845	·9842	·9839	·9836	·9833	·9829	·9826	·9823	·9820	1	1	2	2	3
11	0·9816	·9813	·9810	·9806	·9803	·9799	·9796	·9792	·9789	·9785	1	1	2	2	3
12	0·9781	·9778	·9774	·9770	·9767	·9763	·9759	·9755	·9751	·9748	1	1	2	2	3
13	0·9744	·9740	·9736	·9732	·9728	·9724	·9720	·9715	·9711	·9707	1	1	2	3	3
14	0·9703	·9699	·9694	·9690	·9686	·9681	·9677	·9673	·9668	·9664	1	1	2	3	4
15	0·9659	·9655	·9650	·9646	·9641	·9636	·9632	·9627	·9622	·9617	1	2	2	3	4
16	0·9613	·9608	·9603	·9598	·9593	·9588	·9583	·9578	·9573	·9568	1	2	2	3	4
17	0·9563	·9558	·9553	·9548	·9542	·9537	·9532	·9527	·9521	·9516	1	2	3	3	4
18	0·9511	·9505	·9500	·9494	·9489	·9483	·9478	·9472	·9466	·9461	1	2	3	4	5
19	0·9455	·9449	·9444	·9438	·9432	·9426	·9421	·9415	·9409	·9403	1	2	3	4	5
20	0·9397	·9391	·9385	·9379	·9373	·9367	·9361	·9354	·9348	·9342	1	2	3	4	5
21	0·9336	·9330	·9323	·9317	·9311	·9304	·9298	·9291	·9285	·9278	1	2	3	4	5
22	0·9272	·9265	·9259	·9252	·9245	·9239	·9232	·9225	·9219	·9212	1	2	3	4	6
23	0·9205	·9198	·9191	·9184	·9178	·9171	·9164	·9157	·9150	·9143	1	2	3	5	6
24	0·9135	·9128	·9121	·9114	·9107	·9100	·9092	·9085	·9078	·9070	1	2	4	5	6
25	0·9063	·9056	·9048	·9041	·9033	·9026	·9018	·9011	·9003	·8996	1	2	4	5	6
26	0·8988	·8980	·8973	·8965	·8957	·8949	·8942	·8934	·8926	·8918	1	3	4	5	6
27	0·8910	·8902	·8894	·8886	·8878	·8870	·8862	·8854	·8846	·8838	1	3	4	5	7
28	0·8829	·8821	·8813	·8805	·8796	·8788	·8780	·8771	·8763	·8755	1	3	4	6	7
29	0·8746	·8738	·8729	·8721	·8712	·8704	·8695	·8686	·8678	·8669	1	3	4	6	7
30	0·8660	·8652	·8643	·8634	·8625	·8616	·8607	·8599	·8590	·8581	1	3	4	6	7
31	0·8572	·8563	·8554	·8545	·8536	·8526	·8517	·8508	·8499	·8490	2	3	5	6	8
32	0·8480	·8471	·8462	·8453	·8443	·8434	·8425	·8415	·8406	·8396	2	3	5	6	8
33	0·8387	·8377	·8368	·8358	·8348	·8339	·8329	·8320	·8310	·8300	2	3	5	6	8
34	0·8290	·8281	·8271	·8261	·8251	·8241	·8231	·8221	·8211	·8202	2	3	5	7	8
35	0·8192	·8181	·8171	·8161	·8151	·8141	·8131	·8121	·8111	·8100	2	3	5	7	8
36	0·8090	·8080	·8070	·8059	·8049	·8039	·8028	·8018	·8007	·7997	2	3	5	7	9
37	0·7986	·7976	·7965	·7955	·7944	·7934	·7923	·7912	·7902	·7891	2	4	5	7	9
38	0·7880	·7869	·7859	·7848	·7837	·7826	·7815	·7804	·7793	·7782	2	4	5	7	9
39	0·7771	·7760	·7749	·7738	·7727	·7716	·7705	·7694	·7683	·7672	2	4	6	7	9
40	0·7660	·7649	·7638	·7627	·7615	·7604	·7593	·7581	·7570	·7559	2	4	6	8	9
41	0·7547	·7536	·7524	·7513	·7501	·7490	·7478	·7466	·7455	·7443	2	4	6	8	10
42	0·7431	·7420	·7408	·7396	·7385	·7373	·7361	·7349	·7337	·7325	2	4	6	8	10
43	0·7314	·7302	·7290	·7278	·7266	·7254	·7242	·7230	·7218	·7206	2	4	6	8	10
44	0·7193	·7181	·7169	·7157	·7145	·7133	·7120	·7108	·7096	·7083	2	4	6	8	10
	0′	6′	12′	18′	24′	30′	36′	42′	48′	54′	1′	2′	3′	4′	5′

Proportional
Parts
Subtract

	0′	6′	12′	18′	24′	30′	36′	42′	48′	54′	1′	2′	3′	4′	5′
45°	0·7071	·7059	·7046	·7034	·7022	·7009	·6997	·6984	·6972	·6959	2	4	6	8	10
46	0·6947	·6934	·6921	·6909	·6896	·6884	·6871	·6858	·6845	·6833	2	4	6	8	11
47	0·6820	·6807	·6794	·6782	·6769	·6756	·6743	·6730	·6717	·6704	2	4	6	9	11
48	0·6691	·6678	·6665	·6652	·6639	·6626	·6613	·6600	·6587	·6574	2	4	6	9	11
49	0·6561	·6547	·6534	·6521	·6508	·6494	·6481	·6468	·6455	·6441	2	4	7	9	11
50	0·6428	·6414	·6401	·6388	·6374	·6361	·6347	·6334	·6320	·6307	2	4	7	9	11
51	0·6293	·6280	·6266	·6252	·6239	·6225	·6211	·6198	·6184	·6170	2	5	7	9	11
52	0·6157	·6143	·6129	·6115	·6101	·6088	·6074	·6060	·6046	·6032	2	5	7	9	12
53	0·6018	·6004	·5990	·5976	·5962	·5948	·5934	·5920	·5906	·5892	2	5	7	9	12
54	0·5878	·5864	·5850	·5835	·5821	·5807	·5793	·5779	·5764	·5750	2	5	7	9	12
55	0·5736	·5721	·5707	·5693	·5678	·5664	·5650	·5635	·5621	·5606	2	5	7	10	12
56	0·5592	·5577	·5563	·5548	·5534	·5519	·5505	·5490	·5476	·5461	2	5	7	10	12
57	0·5446	·5432	·5417	·5402	·5388	·5373	·5358	·5344	·5329	·5314	2	5	7	10	12
58	0·5299	·5284	·5270	·5255	·5240	·5225	·5210	·5195	·5180	·5165	2	5	7	10	12
59	0·5150	·5135	·5120	·5105	·5090	·5075	·5060	·5045	·5030	·5015	2	5	8	10	12
60	0·5000	·4985	·4970	·4955	·4939	·4924	·4909	·4894	·4879	·4863	3	5	8	10	13
61	0·4848	·4833	·4818	·4802	·4787	·4772	·4756	·4741	·4726	·4710	3	5	8	10	13
62	0·4695	·4679	·4664	·4648	·4633	·4617	·4602	·4586	·4571	·4555	3	5	8	10	13
63	0·4540	·4524	·4509	·4493	·4478	·4462	·4446	·4431	·4415	·4399	3	5	8	10	13
64	0·4384	·4368	·4352	·4337	·4321	·4305	·4289	·4274	·4258	·4242	3	5	8	11	13
65	0·4226	·4210	·4195	·4179	·4163	·4147	·4131	·4115	·4099	·4083	3	5	8	11	13
66	0·4067	·4051	·4035	·4019	·4003	·3987	·3971	·3955	·3939	·3923	3	5	8	11	13
67	0·3907	·3891	·3875	·3859	·3843	·3827	·3811	·3795	·3778	·3762	3	5	8	11	13
68	0·3746	·3730	·3714	·3697	·3681	·3665	·3649	·3633	·3616	·3600	3	5	8	11	14
69	0·3584	·3567	·3551	·3535	·3518	·3502	·3486	·3469	·3453	·3437	3	5	8	11	14
70	0·3420	·3404	·3387	·3371	·3355	·3338	·3322	·3305	·3289	·3272	3	5	8	11	14
71	0·3256	·3239	·3223	·3206	·3190	·3173	·3156	·3140	·3123	·3107	3	6	8	11	14
72	0·3090	·3074	·3057	·3040	·3024	·3007	·2990	·2974	·2957	·2940	3	6	8	11	14
73	0·2924	·2907	·2890	·2874	·2857	·2840	·2823	·2807	·2790	·2773	3	6	8	11	14
74	0·2756	·2740	·2723	·2706	·2689	·2672	·2656	·2639	·2622	·2605	3	6	8	11	14
75	0·2588	·2571	·2554	·2538	·2521	·2504	·2487	·2470	·2453	·2436	3	6	8	11	14
76	0·2419	·2402	·2385	·2368	·2351	·2334	·2317	·2300	·2284	·2267	3	6	8	11	14
77	0·2250	·2232	·2215	·2198	·2181	·2164	·2147	·2130	·2113	·2096	3	6	9	11	14
78	0·2079	·2062	·2045	·2028	·2011	·1994	·1977	·1959	·1942	·1925	3	6	9	11	14
79	0·1908	·1891	·1874	·1857	·1840	·1822	·1805	·1788	·1771	·1754	3	6	9	11	14
80	0·1736	·1719	·1702	·1685	·1668	·1650	·1633	·1616	·1599	·1582	3	6	9	11	14
81	0·1564	·1547	·1530	·1513	·1495	·1478	·1461	·1444	·1426	·1409	3	6	9	11	14
82	0·1392	·1374	·1357	·1340	·1323	·1305	·1288	·1271	·1253	·1236	3	6	9	12	14
83	0·1219	·1201	·1184	·1167	·1149	·1132	·1115	·1097	·1080	·1063	3	6	9	12	14
84	0·1045	·1028	·1011	·0993	·0976	·0958	·0941	·0924	·0906	·0889	3	6	9	12	14
85	0·0872	·0854	·0837	·0819	·0802	·0785	·0767	·0750	·0732	·0715	3	6	9	12	14
86	0·0698	·0680	·0663	·0645	·0628	·0610	·0593	·0576	·0558	·0541	3	6	9	12	15
87	0·0523	·0506	·0489	·0471	·0454	·0436	·0419	·0401	·0384	·0366	3	6	9	12	15
88	0·0349	·0332	·0314	·0297	·0279	·0262	·0244	·0227	·0209	·0192	3	6	9	12	15
89	0·0175	·0157	·0140	·0122	·0105	·0087	·0070	·0052	·0035	·0017	3	6	9	12	15
	0′	6′	12′	18′	24′	30′	36′	42′	48′	54′	1′	2′	3′	4′	5′

NATURAL TANGENTS

	0'	6'	12'	18'	24'	30'	36'	42'	48'	54'	1'	2'	3'	4'	5'
0°	0·0000	·0017	·0035	·0052	·0070	·0087	·0105	·0122	·0140	·0157	3	6	9	12	15
1	0·0175	·0192	·0209	·0227	·0244	·0262	·0279	·0297	·0314	·0332	3	6	9	12	15
2	0·0349	·0367	·0384	·0402	·0419	·0437	·0454	·0472	·0489	·0507	3	6	9	12	15
3	0·0524	·0542	·0559	·0577	·0594	·0612	·0629	·0647	·0664	·0682	3	6	9	12	15
4	0·0699	·0717	·0734	·0752	·0769	·0787	·0805	·0822	·0840	·0857	3	6	9	12	15
5	0·0875	·0892	·0910	·0928	·0945	·0963	·0981	·0998	·1016	·1033	3	6	9	12	15
6	0·1051	·1069	·1086	·1104	·1122	·1139	·1157	·1175	·1192	·1210	3	6	9	12	15
7	0·1228	·1246	·1263	·1281	·1299	·1317	·1334	·1352	·1370	·1388	3	6	9	12	15
8	0·1405	·1423	·1441	·1459	·1477	·1495	·1512	·1530	·1548	·1566	3	6	9	12	15
9	0·1584	·1602	·1620	·1638	·1655	·1673	·1691	·1709	·1727	·1745	3	6	9	12	15
10	0·1763	·1781	·1799	·1817	·1835	·1853	·1871	·1890	·1908	·1926	3	6	9	12	15
11	0·1944	·1962	·1980	·1998	·2016	·2035	·2053	·2071	·2089	·2107	3	6	9	12	15
12	0·2126	·2144	·2162	·2180	·2199	·2217	·2235	·2254	·2272	·2290	3	6	9	12	15
13	0·2309	·2327	·2345	·2364	·2382	·2401	·2419	·2438	·2456	·2475	3	6	9	12	15
14	0·2493	·2512	·2530	·2549	·2568	·2586	·2605	·2623	·2642	·2661	3	6	9	12	16
15	0·2679	·2698	·2717	·2736	·2754	·2773	·2792	·2811	·2830	·2849	3	6	9	13	16
16	0·2867	·2886	·2905	·2924	·2943	·2962	·2981	·3000	·3019	·3038	3	6	9	13	16
17	0·3057	·3076	·3096	·3115	·3134	·3153	·3172	·3191	·3211	·3230	3	6	9	13	16
18	0·3249	·3269	·3288	·3307	·3327	·3346	·3365	·3385	·3404	·3424	3	6	10	13	16
19	0·3443	·3463	·3482	·3502	·3522	·3541	·3561	·3581	·3600	·3620	3	6	10	13	16
20	0·3640	·3659	·3679	·3699	·3719	·3739	·3759	·3779	·3799	·3819	3	6	10	13	17
21	0·3839	·3859	·3879	·3899	·3919	·3939	·3959	·3979	·4000	·4020	3	7	10	13	17
22	0·4040	·4061	·4081	·4101	·4122	·4142	·4163	·4183	·4204	·4224	3	7	10	14	17
23	0·4245	·4265	·4286	·4307	·4327	·4348	·4369	·4390	·4411	·4431	3	7	10	14	17
24	0·4452	·4473	·4494	·4515	·4536	·4557	·4578	·4599	·4621	·4642	4	7	11	14	18
25	0·4663	·4684	·4706	·4727	·4748	·4770	·4791	·4813	·4834	·4856	4	7	11	14	18
26	0·4877	·4899	·4921	·4942	·4964	·4986	·5008	·5029	·5051	·5073	4	7	11	15	18
27	0·5095	·5117	·5139	·5161	·5184	·5206	·5228	·5250	·5272	·5295	4	7	11	15	18
28	0·5317	·5339	·5362	·5384	·5407	·5430	·5452	·5475	·5498	·5520	4	8	11	15	19
29	0·5543	·5566	·5589	·5612	·5635	·5658	·5681	·5704	·5727	·5750	4	8	12	15	19
30	0·5774	·5797	·5820	·5844	·5867	·5891	·5914	·5938	·5961	·5985	4	8	12	16	20
31	0·6009	·6032	·6056	·6080	·6104	·6128	·6152	·6176	·6200	·6224	4	8	12	16	20
32	0·6249	·6273	·6297	·6322	·6346	·6371	·6395	·6420	·6445	·6469	4	8	12	16	20
33	0·6494	·6519	·6544	·6569	·6594	·6619	·6644	·6669	·6694	·6720	4	8	13	17	21
34	0·6745	·6771	·6796	·6822	·6847	·6873	·6899	·6924	·6950	·6976	4	9	13	17	21
35	0·7002	·7028	·7054	·7080	·7107	·7133	·7159	·7186	·7212	·7239	4	9	13	18	22
36	0·7265	·7292	·7319	·7346	·7373	·7400	·7427	·7454	·7481	·7508	5	9	14	18	23
37	0·7536	·7563	·7590	·7618	·7646	·7673	·7701	·7729	·7757	·7785	5	9	14	18	23
38	0·7813	·7841	·7869	·7898	·7926	·7954	·7983	·8012	·8040	·8069	5	10	14	19	24
39	0·8098	·8127	·8156	·8185	·8214	·8243	·8273	·8302	·8332	·8361	5	10	15	20	24
40	0·8391	·8421	·8451	·8481	·8511	·8541	·8571	·8601	·8632	·8662	5	10	15	20	25
41	0·8693	·8724	·8754	·8785	·8816	·8847	·8878	·8910	·8941	·8972	5	10	16	21	26
42	0·9004	·9036	·9067	·9099	·9131	·9163	·9195	·9228	·9260	·9293	5	11	16	21	26
43	0·9325	·9358	·9391	·9424	·9457	·9490	·9523	·9556	·9590	·9623	6	11	17	22	28
44	0·9657	·9691	·9725	·9759	·9793	·9827	·9861	·9896	·9930	·9965	6	11	17	23	29
	0'	6'	12'	18'	24'	30'	36'	42'	48'	54'	1'	2'	3'	4'	5'

	0′	6′	12′	18′	24′	30′	36′	42′	48′	54′	1′	2′	3′	4′	5′
45°	1·0000	·0035	·0070	·0105	·0141	·0176	·0212	·0247	·0283	·0319	6	12	18	24	30
46	1·0355	·0392	·0428	·0464	·0501	·0538	·0575	·0612	·0649	·0686	6	12	18	25	31
47	1·0724	·0761	·0799	·0837	·0875	·0913	·0951	·0990	·1028	·1067	6	13	19	25	32
48	1·1106	·1145	·1184	·1224	·1263	·1303	·1343	·1383	·1423	·1463	7	13	20	27	33
49	1·1504	·1544	·1585	·1626	·1667	·1708	·1750	·1792	·1833	·1875	7	14	21	28	34
50	1·1918	·1960	·2002	·2045	·2088	·2131	·2174	·2218	·2261	·2305	7	14	22	29	36
51	1·2349	·2393	·2437	·2482	·2527	·2572	·2617	·2662	·2708	·2753	8	15	23	30	38
52	1·2799	·2846	·2892	·2938	·2985	·3032	·3079	·3127	·3175	·3222	8	16	24	31	39
53	1·3270	·3319	·3367	·3416	·3465	·3514	·3564	·3613	·3663	·3713	8	16	25	33	41
54	1·3764	·3814	·3865	·3916	·3968	·4019	·4071	·4124	·4176	·4229	9	17	26	34	43
55	1·4281	·4335	·4388	·4442	·4496	·4550	·4605	·4659	·4715	·4770	9	18	27	36	45
56	1·4826	·4882	·4938	·4994	·5051	·5108	·5166	·5224	·5282	·5340	10	19	29	38	48
57	1·5399	·5458	·5517	·5577	·5637	·5697	·5757	·5818	·5880	·5941	10	20	30	40	50
58	1·6003	·6066	·6128	·6191	·6255	·6319	·6383	·6447	·6512	·6577	11	21	32	43	53
59	1·6643	·6709	·6775	·6842	·6909	·6977	·7045	·7113	·7182	·7251	11	23	34	45	57
60	1·7321	·7391	·7461	·7532	·7603	·7675	·7747	·7820	·7893	·7966	12	24	36	48	60
61	1·8040	·8115	·8190	·8265	·8341	·8418	·8495	·8572	·8650	·8728	13	26	38	51	64
62	1·8807	·8887	·8967	·9047	·9128	·9210	·9292	·9375	·9458	·9542	14	27	41	55	68
63	1·9626	·9711	·9797	·9883	1·9970	2·0057	·0145	·0233	·0323	·0413	15	29	44	58	73
64	2·0503	·0594	·0686	·0778	·0872	·0965	·1060	·1155	·1251	·1348	16	31	47	63	78
65	2·145	·154	·164	·174	·184	·194	·204	·215	·225	·236	2	3	5	7	8
66	2·246	·257	·267	·278	·289	·300	·311	·322	·333	·344	2	4	5	7	9
67	2·356	·367	·379	·391	·402	·414	·426	·438	·450	·463	2	4	6	8	10
68	2·475	·488	·500	·513	·526	·539	·552	·565	·578	·592	2	4	6	9	11
69	2·605	·619	·633	·646	·660	·675	·689	·703	·718	·733	2	5	7	9	12
70	2·747	·762	·778	·793	·808	·824	·840	·856	·872	·888	3	5	8	10	13
71	2·904	·921	·937	·954	·971	2·989	3·006	·024	·042	·060	3	6	9	12	14
72	3·078	·096	·115	·133	·152	·172	·191	·211	·230	·251	3	6	10	13	16
73	3·271	·291	·312	·333	·354	·376	·398	·420	·442	·465	4	7	11	14	18
74	3·487	·511	·534	·558	·582	·606	·630	·655	·681	·706	4	8	12	16	20
75	3·732	·758	·785	·812	·839	·867	·895	·923	·952	·981	5	9	14	19	23
76	4·011	·041	·071	·102	·134	·165	·198	·230	·264	·297	5	11	16	21	27
77	4·331	·366	·402	·437	·474	·511	·548	·586	·625	·665	6	12	19	25	31
78	4·705	·745	·787	·829	·872	·915	4·959	5·005	·050	·097	7	15	22	29	37
79	5·145	·193	·242	·292	·343	·396	·449	·503	·558	·614	9	18	26	35	44
80	5·671	·730	·789	·850	·912	5·976	6·041	·107	·174	·243	11	21	32	43	54
81	6·314	·386	·460	·535	·612	·691	·772	·855	6·940	7·026	13	27	40	54	67
82	7·115	·207	·300	·396	·495	·596	·700	·806	7·916	8·028	17	34	51	69	86
83	8·144	·264	·386	·513	·643	·777	8·915	9·058	·205	·357	23	46	68	91	114
84	9·514	9·677	9·845	10·019	10·199	10·385	10·579	10·780	10·988	11·205					
85	11·43	11·66	11·91	12·16	12·43	12·71	13·00	13·30	13·62	13·95	p.p. cease				
86	14·30	14·67	15·06	15·46	15·89	16·35	16·83	17·34	17·89	18·46	to be				
87	19·08	19·74	20·45	21·20	22·02	22·90	23·86	24·90	26·03	27·27	sufficiently				
88	28·64	30·14	31·82	33·69	35·80	38·19	40·92	44·07	47·74	52·08	accurate				
89	57·29	63·66	71·62	81·85	95·49	114·6	143·2	191·0	286·5	573·0					
	0′	6′	12′	18′	24′	30′	36′	42′	48′	54′	1′	2′	3′	4′	5′

NATURAL COSECANTS

Proportional Parts Subtract

	0′	6′	12′	18′	24′	30′	36′	42′	48′	54′	1′	2′	3′	4′	5′
0°	∞	573·0	286·5	191·0	143·2	114·6	95·49	81·85	71·62	63·66					
1	57·30	52·09	47·75	44·08	40·93	38·20	35·81	33·71	31·84	30·16					
2	28·65	27·29	26·05	24·92	23·88	22·93	22·04	21·23	20·47	19·77					
3	19·11	18·49	17·91	17·37	16·86	16·38	15·93	15·50	15·09	14·70					
4	14·34	13·99	13·65	13·34	13·03	12·75	12·47	12·20	11·95	11·71	p.p. cease		to be		
5	11·474	·249	11·034	10·826	·626	·433	·248	10·068	9·895	·728	sufficiently				
6	9·567	·411	·259	9·113	8·971	·834	·700	·571	·446	·324	accurate				
7	8·206	8·091	7·979	·870	·764	·661	·561	·463	·368	·276					
8	7·185	·097	7·011	6·927	·845	·765	·687	·611	·537	·464					
9	6·392	·323	·255	·188	·123	6·059	5·996	·935	·875	·816					
10	5·759	·702	·647	·593	·540	·487	·436	·386	·337	·288	9	17	26	35	43
11	5·241	·194	·148	·103	·059	5·016	4·973	·931	·890	·850	7	14	22	29	36
12	4·810	·771	·732	·694	·657	·620	·584	·549	·514	·479	6	12	18	24	30
13	4·445	·412	·379	·347	·315	·284	·253	·222	·192	·163	5	10	16	21	26
14	4·134	·105	·077	·049	4·021	3·994	·967	·941	·915	·889	4	9	14	18	22
15	3·864	·839	·814	·790	·766	·742	·719	·695	·673	·650	4	8	12	16	20
16	3·628	·606	·584	·563	·542	·521	·500	·480	·460	·440	3	7	10	14	17
17	3·420	·401	·382	·363	·344	·326	·307	·289	·271	·254	3	6	9	12	15
18	3·236	·219	·202	·185	·168	·152	·135	·119	·103	·087	3	5	8	11	14
19	3·072	·056	·041	·026	3·011	2·996	·981	·967	·952	·938	2	5	7	10	12
20	2·924	·910	·896	·882	·869	·855	·842	·829	·816	·803	2	4	7	9	11
21	2·790	·778	·765	·753	·741	·729	·716	·705	·693	·681	2	4	6	8	10
22	2·669	·658	·647	·635	·624	·613	·602	·591	·581	·570	2	4	6	7	9
23	2·559	·549	·538	·528	·518	·508	·498	·488	·478	·468	2	3	5	7	8
24	2·459	·449	·439	·430	·421	·411	·402	·393	·384	·375	2	3	5	6	8
25	2·366	·357	·349	·340	·331	·323	·314	·306	·298	·289	1	3	4	6	7
26	2·281	·273	·265	·257	·249	·241	·233	·226	·218	·210	1	3	4	5	7
27	2·203	·195	·188	·180	·173	·166	·158	·151	·144	·137	1	2	4	5	6
28	2·130	·123	·116	·109	·103	·096	·089	·082	·076	·069	1	2	3	4	6
29	2·063	·056	·050	·043	·037	·031	·025	·018	·012	·006	1	2	3	4	5
30	2·0000	1·9940	·9880	·9821	·9762	·9703	·9645	·9587	·9530	·9473	10	19	29	39	49
31	1·9416	·9360	·9304	·9249	·9194	·9139	·9084	·9031	·8977	·8924	9	18	27	36	45
32	1·8871	·8818	·8766	·8714	·8663	·8612	·8561	·8510	·8460	·8410	8	17	26	34	42
33	1·8361	·8312	·8263	·8214	·8166	·8118	·8070	·8023	·7976	·7929	8	16	24	32	40
34	1·7883	·7837	·7791	·7745	·7700	·7655	·7610	·7566	·7522	·7478	7	15	22	30	37
35	1·7434	·7391	·7348	·7305	·7263	·7221	·7179	·7137	·7095	·7054	7	14	21	28	35
36	1·7013	·6972	·6932	·6892	·6852	·6812	·6772	·6733	·6694	·6655	7	13	20	26	33
37	1·6616	·6578	·6540	·6502	·6464	·6427	·6390	·6353	·6316	·6279	6	12	19	25	31
38	1·6243	·6207	·6171	·6135	·6099	·6064	·6029	·5994	·5959	·5925	6	12	18	24	29
39	1·5890	·5856	·5822	·5788	·5755	·5721	·5688	·5655	·5622	·5590	6	11	17	22	28
40	1·5557	·5525	·5493	·5461	·5429	·5398	·5366	·5335	·5304	·5273	5	10	16	21	26
41	1·5243	·5212	·5182	·5151	·5121	·5092	·5062	·5032	·5003	·4974	5	10	15	20	25
42	1·4945	·4916	·4887	·4859	·4830	·4802	·4774	·4746	·4718	·4690	5	9	14	19	24
43	1·4663	·4635	·4608	·4581	·4554	·4527	·4501	·4474	·4448	·4422	4	9	13	18	22
44	1·4396	·4370	·4344	·4318	·4293	·4267	·4242	·4217	·4192	·4167	4	8	13	17	21
	0′	6′	12′	18′	24′	30′	36′	42′	48′	54′	1′	2′	3′	4′	5′

NATURAL COSECANTS

Proportional
Parts
Subtract

	0′	6′	12′	18′	24′	30′	36′	42′	48′	54′	1′	2′	3′	4′	5′
45°	1·4142	·4118	·4093	·4069	·4044	·4020	·3996	·3972	·3949	·3925	4	8	12	16	20
46	1·3902	·3878	·3855	·3832	·3809	·3786	·3763	·3741	·3718	·3696	4	8	11	15	19
47	1·3673	·3651	·3629	·3607	·3585	·3563	·3542	·3520	·3499	·3478	4	7	11	14	18
48	1·3456	·3435	·3414	·3393	·3373	·3352	·3331	·3311	·3291	·3270	3	7	10	14	17
49	1·3250	·3230	·3210	·3190	·3171	·3151	·3131	·3112	·3093	·3073	3	7	10	13	16
50	1·3054	·3035	·3016	·2997	·2978	·2960	·2941	·2923	·2904	·2886	3	6	9	12	15
51	1·2868	·2849	·2831	·2813	·2796	·2778	·2760	·2742	·2725	·2708	3	6	9	12	15
52	1·2690	·2673	·2656	·2639	·2622	·2605	·2588	·2571	·2554	·2538	3	6	8	11	14
53	1·2521	·2505	·2489	·2472	·2456	·2440	·2424	·2408	·2392	·2376	3	5	8	11	13
54	1·2361	·2345	·2329	·2314	·2299	·2283	·2268	·2253	·2238	·2223	3	5	8	10	13
55	1·2208	·2193	·2178	·2163	·2149	·2134	·2120	·2105	·2091	·2076	2	5	7	10	12
56	1·2062	·2048	·2034	·2020	·2006	·1992	·1978	·1964	·1951	·1937	2	5	7	9	12
57	1·1924	·1910	·1897	·1883	·1870	·1857	·1844	·1831	·1818	·1805	2	4	7	9	11
58	1·1792	·1779	·1766	·1753	·1741	·1728	·1716	·1703	·1691	·1679	2	4	6	8	10
59	1·1666	·1654	·1642	·1630	·1618	·1606	·1594	·1582	·1570	·1559	2	4	6	8	10
60	1·1547	·1535	·1524	·1512	·1501	·1490	·1478	·1467	·1456	·1445	2	4	6	8	9
61	1·1434	·1423	·1412	·1401	·1390	·1379	·1368	·1357	·1347	·1336	2	4	5	7	9
62	1·1326	·1315	·1305	·1294	·1284	·1274	·1264	·1253	·1243	·1233	2	3	5	7	9
63	1·1223	·1213	·1203	·1194	·1184	·1174	·1164	·1155	·1145	·1136	2	3	5	6	8
64	1·1126	·1117	·1107	·1098	·1089	·1079	·1070	·1061	·1052	·1043	2	3	5	6	8
65	1·1034	·1025	·1016	·1007	·0998	·0989	·0981	·0972	·0963	·0955	1	3	4	6	7
66	1·0946	·0938	·0929	·0921	·0913	·0904	·0896	·0888	·0880	·0872	1	3	4	5	7
67	1·0864	·0856	·0848	·0840	·0832	·0824	·0816	·0808	·0801	·0793	1	3	4	5	7
68	1·0785	·0778	·0770	·0763	·0755	·0748	·0740	·0733	·0726	·0719	1	2	4	5	6
69	1·0711	·0704	·0697	·0690	·0683	·0676	·0669	·0662	·0655	·0649	1	2	3	5	6
70	1·0642	·0635	·0628	·0622	·0615	·0608	·0602	·0595	·0589	·0583	1	2	3	4	5
71	1·0576	·0570	·0564	·0557	·0551	·0545	·0539	·0533	·0527	·0521	1	2	3	4	5
72	1·0515	·0509	·0503	·0497	·0491	·0485	·0480	·0474	·0468	·0463	1	2	3	4	5
73	1·0457	·0451	·0446	·0440	·0435	·0429	·0424	·0419	·0413	·0408	1	2	3	4	5
74	1·0403	·0398	·0393	·0388	·0382	·0377	·0372	·0367	·0363	·0358	1	2	3	3	4
75	1·0353	·0348	·0343	·0338	·0334	·0329	·0324	·0320	·0315	·0311	1	2	2	3	4
76	1·0306	·0302	·0297	·0293	·0288	·0284	·0280	·0276	·0271	·0267	1	1	2	3	4
77	1·0263	·0259	·0255	·0251	·0247	·0243	·0239	·0235	·0231	·0227	1	1	2	3	3
78	1·0223	·0220	·0216	·0212	·0209	·0205	·0201	·0198	·0194	·0191	1	1	2	2	3
79	1·0187	·0184	·0180	·0177	·0174	·0170	·0167	·0164	·0161	·0157	1	1	2	2	3
80	1·0154	·0151	·0148	·0145	·0142	·0139	·0136	·0133	·0130	·0127	0	1	1	2	2
81	1·0125	·0122	·0119	·0116	·0114	·0111	·0108	·0106	·0103	·0101	0	1	1	2	2
82	1·0098	·0096	·0093	·0091	·0089	·0086	·0084	·0082	·0079	·0077	0	1	1	2	2
83	1·0075	·0073	·0071	·0069	·0067	·0065	·0063	·0061	·0059	·0057	0	1	1	1	2
84	1·0055	·0053	·0051	·0050	·0048	·0046	·0045	·0043	·0041	·0040	0	1	1	1	1
85	1·0038	·0037	·0035	·0034	·0032	·0031	·0030	·0028	·0027	·0026	0	0	1	1	1
86	1·0024	·0023	·0022	·0021	·0020	·0019	·0018	·0017	·0016	·0015	0	0	0	1	1
87	1·0014	·0013	·0012	·0011	·0010	·0010	·0009	·0008	·0007	·0007	0	0	0	1	1
88	1·0006	·0005	·0005	·0004	·0004	·0003	·0003	·0003	·0002	·0002	0	0	0	0	0
89	1·0002	·0001	·0001	·0001	·0001	·0000	·0000	·0000	·0000	·0000	0	0	0	0	0
	0′	6′	12′	18′	24′	30′	36′	42′	48′	54′	1′	2′	3′	5′	4′

NATURAL SECANTS

	0′	6′	12′	18′	24′	30′	36′	42′	48′	54′	1′	2′	3′	4′	5′
0°	1·0000	·0000	·0000	·0000	·0000	·0000	·0001	·0001	·0001	·0001	0	0	0	0	0
1	1·0002	·0002	·0002	·0003	·0003	·0003	·0004	·0004	·0005	·0005	0	0	0	0	0
2	1·0006	·0007	·0007	·0008	·0009	·0010	·0010	·0011	·0012	·0013	0	0	0	1	1
3	1·0014	·0015	·0016	·0017	·0018	·0019	·0020	·0021	·0022	·0023	0	0	1	1	1
4	1·0024	·0026	·0027	·0028	·0030	·0031	·0032	·0034	·0035	·0037	0	0	1	1	1
5	1·0038	·0040	·0041	·0043	·0045	·0046	·0048	·0050	·0051	·0053	0	1	1	1	1
6	1·0055	·0057	·0059	·0061	·0063	·0065	·0067	·0069	·0071	·0073	0	1	1	1	2
7	1·0075	·0077	·0079	·0082	·0084	·0086	·0089	·0091	·0093	·0096	0	1	1	2	2
8	1·0098	·0101	·0103	·0106	·0108	·0111	·0114	·0116	·0119	·0122	0	1	1	2	2
9	1·0125	·0127	·0130	·0133	·0136	·0139	·0142	·0145	·0148	·0151	0	1	1	2	2
10	1·0154	·0157	·0161	·0164	·0167	·0170	·0174	·0177	·0180	·0184	1	1	2	2	3
11	1·0187	·0191	·0194	·0198	·0201	·0205	·0209	·0212	·0216	·0220	1	1	2	2	3
12	1·0223	·0227	·0231	·0235	·0239	·0243	·0247	·0251	·0255	·0259	1	1	2	3	3
13	1·0263	·0267	·0271	·0276	·0280	·0284	·0288	·0293	·0297	·0302	1	1	2	3	4
14	1·0306	·0311	·0315	·0320	·0324	·0329	·0334	·0338	·0343	·0348	1	2	2	3	4
15	1·0353	·0358	·0363	·0367	·0372	·0377	·0382	·0388	·0393	·0398	1	2	3	3	4
16	1·0403	·0408	·0413	·0419	·0424	·0429	·0435	·0440	·0446	·0451	1	2	3	4	5
17	1·0457	·0463	·0468	·0474	·0480	·0485	·0491	·0497	·0503	·0509	1	2	3	4	5
18	1·0515	·0521	·0527	·0533	·0539	·0545	·0551	·0557	·0564	·0570	1	2	3	4	5
19	1·0576	·0583	·0589	·0595	·0602	·0608	·0615	·0622	·0628	·0635	1	2	3	4	5
20	1·0642	·0649	·0655	·0662	·0669	·0676	·0683	·0690	·0697	·0704	1	2	3	5	6
21	1·0711	·0719	·0726	·0733	·0740	·0748	·0755	·0763	·0770	·0778	1	2	4	5	6
22	1·0785	·0793	·0801	·0808	·0816	·0824	·0832	·0840	·0848	·0856	1	3	4	5	7
23	1·0864	·0872	·0880	·0888	·0896	·0904	·0913	·0921	·0929	·0938	1	3	4	5	7
24	1·0946	·0955	·0963	·0972	·0981	·0989	·0998	·1007	·1016	·1025	1	3	4	6	7
25	1·1034	·1043	·1052	·1061	·1070	·1079	·1089	·1098	·1107	·1117	2	3	5	6	8
26	1·1126	·1136	·1145	·1155	·1164	·1174	·1184	·1194	·1203	·1213	2	3	5	6	8
27	1·1223	·1233	·1243	·1253	·1264	·1274	·1284	·1294	·1305	·1315	2	3	5	7	9
28	1·1326	·1336	·1347	·1357	·1368	·1379	·1390	·1401	·1412	·1423	2	4	5	7	9
29	1·1434	·1445	·1456	·1467	·1478	·1490	·1501	·1512	·1524	·1535	2	4	6	8	9
30	1·1547	·1559	·1570	·1582	·1594	·1606	·1618	·1630	·1642	·1654	2	4	6	8	10
31	1·1666	·1679	·1691	·1703	·1716	·1728	·1741	·1753	·1766	·1779	2	4	6	8	10
32	1·1792	·1805	·1818	·1831	·1844	·1857	·1870	·1883	·1897	·1910	2	4	7	9	11
33	1·1924	·1937	·1951	·1964	·1978	·1992	·2006	·2020	·2034	·2048	2	5	7	9	12
34	1·2062	·2076	·2091	·2105	·2120	·2134	·2149	·2163	·2178	·2193	2	5	7	10	12
35	1·2208	·2223	·2238	·2253	·2268	·2283	·2299	·2314	·2329	·2345	3	5	8	10	13
36	1·2361	·2376	·2392	·2408	·2424	·2440	·2456	·2472	·2489	·2505	3	5	8	11	13
37	1·2521	·2538	·2554	·2571	·2588	·2605	·2622	·2639	·2656	·2673	3	6	8	11	14
38	1·2690	·2708	·2725	·2742	·2760	·2778	·2796	·2813	·2831	·2849	3	6	9	12	15
39	1·2868	·2886	·2904	·2923	·2941	·2960	·2978	·2997	·3016	·3035	3	6	9	12	15
40	1·3054	·3073	·3093	·3112	·3131	·3151	·3171	·3190	·3210	·3230	3	7	10	13	16
41	1·3250	·3270	·3291	·3311	·3331	·3352	·3373	·3393	·3414	·3435	3	7	10	14	17
42	1·3456	·3478	·3499	·3520	·3542	·3563	·3585	·3607	·3629	·3651	4	7	11	14	18
43	1·3673	·3696	·3718	·3741	·3763	·3786	·3809	·3832	·3855	·3878	4	8	11	15	19
44	1·3902	·3925	·3949	·3972	·3996	·4020	·4044	·4069	·4093	·4118	4	8	12	16	20
	0′	6′	12′	18′	24′	30′	36′	42′	48′	54′	1′	2′	3′	4′	5′

NATURAL SECANTS

	0'	6'	12'	18'	24'	30'	36'	42'	48'	54'	1'	2'	3'	4'	5'
45°	1·4142	·4167	·4192	·4217	·4242	·4267	·4293	·4318	·4344	·4370	4	8	13	17	21
46	1·4396	·4422	·4448	·4474	·4501	·4527	·4554	·4581	·4608	·4635	4	9	13	18	22
47	1·4663	·4690	·4718	·4746	·4774	·4802	·4830	·4859	·4887	·4916	5	9	14	19	24
48	1·4945	·4974	·5003	·5032	·5062	·5092	·5121	·5151	·5182	·5212	5	10	15	20	25
49	1·5243	·5273	·5304	·5335	·5366	·5398	·5429	·5461	·5493	·5525	5	10	16	21	26
50	1·5557	·5590	·5622	·5655	·5688	·5721	·5755	·5788	·5822	·5856	6	11	17	22	28
51	1·5890	·5925	·5959	·5994	·6029	·6064	·6099	·6135	·6171	·6207	6	12	18	24	29
52	1·6243	·6279	·6316	·6353	·6390	·6427	·6464	·6502	·6540	·6578	6	12	19	25	31
53	1·6616	·6655	·6694	·6733	·6772	·6812	·6852	·6892	·6932	·6972	7	13	20	26	33
54	1·7013	·7054	·7095	·7137	·7179	·7221	·7263	·7305	·7348	·7391	7	14	21	28	35
55	1·7434	·7478	·7522	·7566	·7610	·7655	·7700	·7745	·7791	·7837	7	15	22	30	37
56	1·7883	·7929	·7976	·8023	·8070	·8118	·8166	·8214	·8263	·8312	8	16	24	32	40
57	1·8361	·8410	·8460	·8510	·8561	·8612	·8663	·8714	·8766	·8818	8	17	26	34	42
58	1·8871	·8924	·8977	·9031	·9084	·9139	·9194	·9249	·9304	·9360	9	18	27	36	45
59	1·9416	·9473	·9530	·9587	·9645	·9703	·9762	·9821	·9880	·9940	10	19	29	39	49
60	2·000	·006	·012	·018	·025	·031	·037	·043	·050	·056	1	2	3	4	5
61	2·063	·069	·076	·082	·089	·096	·103	·109	·116	·123	1	2	3	4	6
62	2·130	·137	·144	·151	·158	·166	·173	·180	·188	·195	1	2	4	5	6
63	2·203	·210	·218	·226	·233	·241	·249	·257	·265	·273	1	3	4	5	7
64	2·281	·289	·298	·306	·314	·323	·331	·340	·349	·357	1	3	4	6	7
65	2·366	·375	·384	·393	·402	·411	·421	·430	·439	·449	2	3	5	6	8
66	2·459	·468	·478	·488	·498	·508	·518	·528	·538	·549	2	3	5	7	8
67	2·559	·570	·581	·591	·602	·613	·624	·635	·647	·658	2	4	6	7	9
68	2·669	·681	·693	·705	·716	·729	·741	·753	·765	·778	2	4	6	8	10
69	2·790	·803	·816	·829	·842	·855	·869	·882	·896	·910	2	4	7	9	11
70	2·924	·938	·952	·967	·981	2·996	3·011	·026	·041	·056	2	5	7	10	12
71	3·072	·087	·103	·119	·135	·152	·168	·185	·202	·219	3	5	8	11	14
72	3·236	·254	·271	·289	·307	·326	·344	·363	·382	·401	3	6	9	12	15
73	3·420	·440	·460	·480	·500	·521	·542	·563	·584	·606	3	7	10	14	17
74	3·628	·650	·673	·695	·719	·742	·766	·790	·814	·839	4	8	12	16	20
75	3·864	·889	·915	·941	·967	3·994	4·021	·049	·077	·105	4	9	14	18	22
76	4·134	·163	·192	·222	·253	·284	·315	·347	·379	·412	5	10	16	21	26
77	4·445	·479	·514	·549	·584	·620	·657	·694	·732	·771	6	12	18	24	30
78	4·810	·850	·890	·931	4·973	5·016	·059	·103	·148	·194	7	14	22	29	36
79	5·241	·288	·337	·386	·436	·487	·540	·593	·647	·702	9	17	26	35	43
80	5·759	·816	·875	·935	5·996	6·059	·123	·188	·255	·323					
81	6·392	·464	·537	·611	·687	·765	·845	6·927	7·011	·097					
82	7·185	·276	·368	·463	·561	·661	·764	·870	7·979	8·091					
83	8·206	·324	·446	·571	·700	·834	8·971	9·113	·259	·411					
84	9·567	·728	9·895	10·068	·248	·433	·626	10·826	11·034	·249					
85	11·47	11·71	11·95	12·20	12·47	12·75	13·03	13·34	13·65	13·99					
86	14·34	14·70	15·09	15·50	15·93	16·38	16·86	17·37	17·91	18·49					
87	19·11	19·77	20·47	21·23	22·04	22·93	23·88	24·92	26·05	27·29					
88	28·65	30·16	31·84	33·71	35·81	38·20	40·93	44·08	47·75	52·09					
89	57·30	63·66	71·62	81·85	95·49	114·6	143·2	191·0	286·5	573·0					
	0'	6'	12'	18'	24'	30'	36'	42'	48'	54'	1'	2'	3'	4'	5'

p.p. cease to be sufficiently accurate

NATURAL COTANGENTS

	0′	6′	12′	18′	24′	30′	36′	42′	48′	54′	1′	2′	3′	4′	5′
0°	∝	573·0	286·5	191·0	143·2	114·6	95·49	81·85	71·62	63·66					
1	57·29	52·08	47·74	44·07	40·92	38·19	35·80	33·69	31·82	30·14		p.p. cease			
2	28·64	27·27	26·03	24·90	23·86	22·90	22·02	21·20	20·45	19·74		to be			
3	19·08	18·46	17·89	17·34	16·83	16·35	15·89	15·46	15·06	14·67		sufficiently			
4	14·30	13·95	13·62	13·30	13·00	12·71	12·43	12·16	11·91	11·66		accurate			
5	11·30	11·205	10·988	10·780	10·579	10·385	10·199	10·019	9·845	9·677					
6	9·514	·357	·205	9·058	8·915	·777	·643	·513	·386	·264	23	46	68	91	114
7	8·144	8·028	7·916	·806	·700	·596	·495	·396	·300	·207	17	34	51	69	86
8	7·115	7·026	6·940	·855	·772	·691	·612	·535	·460	·386	13	27	40	54	67
9	6·314	·243	·174	·107	6·041	5·976	·912	·850	·789	·730	11	21	32	43	54
10	5·671	·614	·558	·503	·449	·396	·343	·292	·242	·193	9	18	26	35	44
11	5·145	·097	·050	5·005	4·959	·915	·872	·829	·787	·745	7	15	22	29	37
12	4·705	·665	·625	·586	·548	·511	·474	·437	·402	·366	6	12	19	25	31
13	4·331	·297	·264	·230	·198	·165	·134	·102	·071	·041	5	11	16	21	27
14	4·011	3·981	·952	·923	·895	·867	·839	·812	·785	·758	5	9	14	19	23
15	3·732	·706	·681	·655	·630	·606	·582	·558	·534	·511	4	8	12	16	20
16	3·487	·465	·442	·420	·398	·376	·354	·333	·312	·291	4	7	11	14	18
17	3·271	·251	·230	·211	·191	·172	·152	·133	·115	·096	3	6	10	13	16
18	3·078	·060	·042	·024	3·006	2·989	·971	·954	·937	·921	3	6	9	12	14
19	2·904	·888	·872	·856	·840	·824	·808	·793	·778	·762	3	5	8	10	13
20	2·747	·733	·718	·703	·689	·675	·660	·646	·633	·619	2	5	7	9	12
21	2·605	·592	·578	·565	·552	·539	·526	·513	·500	·488	2	4	6	9	11
22	2·475	·463	·450	·438	·426	·414	·402	·391	·379	·367	2	4	6	8	10
23	2·356	·344	·333	·322	·311	·300	·289	·278	·267	·257	2	4	5	7	9
24	2·246	·236	·225	·215	·204	·194	·184	·174	·164	·154	2	3	5	7	8
25	2·1445	·1348	·1251	·1155	·1060	·0965	·0872	·0778	·0686	·0594	16	31	47	63	78
26	2·0503	·0413	·0323	·0233	·0145	2·0057	1·9970	·9883	·9797	·9711	15	29	44	58	73
27	1·9626	·9542	·9458	·9375	·9292	·9210	·9128	·9047	·8967	·8887	14	27	41	55	68
28	1·8807	·8728	·8650	·8572	·8495	·8418	·8341	·8265	·8190	·8115	13	26	38	51	64
29	1·8040	·7966	·7893	·7820	·7747	·7675	·7603	·7532	·7461	·7391	12	24	36	48	60
30	1·7321	·7251	·7182	·7113	·7045	·6977	·6909	·6842	·6775	·6709	11	23	34	45	57
31	1·6643	·6577	·6512	·6447	·6383	·6319	·6255	·6191	·6128	·6066	11	21	32	43	53
32	1·6003	·5941	·5880	·5818	·5757	·5697	·5637	·5577	·5517	·5458	10	20	30	40	50
33	1·5399	·5340	·5282	·5224	·5166	·5108	·5051	·4994	·4938	·4882	10	19	29	38	48
34	1·4826	·4770	·4715	·4659	·4605	·4550	·4496	·4442	·4388	·4335	9	18	27	36	45
35	1·4281	·4229	·4176	·4124	·4071	·4019	·3968	·3916	·3865	·3814	9	17	26	34	43
36	1·3764	·3713	·3663	·3613	·3564	·3514	·3465	·3416	·3367	·3319	8	16	25	33	41
37	1·3270	·3222	·3175	·3127	·3079	·3032	·2985	·2938	·2892	·2846	8	16	24	31	39
38	1·2799	·2753	·2708	·2662	·2617	·2572	·2527	·2482	·2437	·2393	8	15	23	30	38
39	1·2349	·2305	·2261	·2218	·2174	·2131	·2088	·2045	·2002	·1960	7	14	22	29	36
40	1·1918	·1875	·1833	·1792	·1750	·1708	·1667	·1626	·1585	·1544	7	14	21	28	34
41	1·1504	·1463	·1423	·1383	·1343	·1303	·1263	·1224	·1184	·1145	7	13	20	27	33
42	1·1106	·1067	·1028	·0990	·0951	·0913	·0875	·0837	·0799	·0761	6	13	19	25	32
43	1·0724	·0686	·0649	·0612	·0575	·0538	·0501	·0464	·0428	·0392	6	12	18	25	31
44	1·0355	·0319	·0283	·0247	·0212	·0176	·0141	·0105	·0070	·0035	6	12	18	24	30
	0′	6′	12	18′	24′	30′	36′	42′	48′	54′	1′	2′	3′	4′	5′

NATURAL COTANGENTS

	0′	6′	12′	18′	24′	30′	36′	42′	48′	54′	1′	2′	3′	4′	5′
45°	1·0000	0·9965	·9930	·9896	·9861	·9827	·9793	·9759	·9725	·9691	6	11	17	23	29
46	0·9657	·9623	·9590	·9556	·9523	·9490	·9457	·9424	·9391	·9358	6	11	17	22	28
47	0·9325	·9293	·9260	·9228	·9195	·9163	·9131	·9099	·9067	·9036	5	11	16	21	27
48	0·9004	·8972	·8941	·8910	·8878	·8847	·8816	·8785	·8754	·8724	5	10	16	21	26
49	0·8693	·8662	·8632	·8601	·8571	·8541	·8511	·8481	·8451	·8421	5	10	15	20	25
50	0·8391	·8361	·8332	·8302	·8273	·8243	·8214	·8185	·8156	·8127	5	10	15	20	24
51	0·8098	·8069	·8040	·8012	·7983	·7954	·7926	·7898	·7869	·7841	5	10	14	19	24
52	0·7813	·7785	·7757	·7729	·7701	·7673	·7646	·7618	·7590	·7563	5	9	14	18	23
53	0·7536	·7508	·7481	·7454	·7427	·7400	·7373	·7346	·7319	·7292	5	9	14	18	23
54	0·7265	·7239	·7212	·7186	·7159	·7133	·7107	·7080	·7054	·7028	4	9	13	18	22
55	0·7002	·6976	·6950	·6924	·6899	·6873	·6847	·6822	·6796	·6771	4	9	13	17	21
56	0·6745	·6720	·6694	·6669	·6644	·6619	·6594	·6569	·6544	·6519	4	8	13	17	21
57	0·6494	·6469	·6445	·6420	·6395	·6371	·6346	·6322	·6297	·6273	4	8	12	16	20
58	0·6249	·6224	·6200	·6176	·6152	·6128	·6104	·6080	·6056	·6032	4	8	12	16	20
59	0·6009	·5985	·5961	·5938	·5914	·5891	·5867	·5844	·5820	·5797	4	8	12	16	20
60	0·5774	·5750	·5727	·5704	·5681	·5658	·5635	·5612	·5589	·5566	4	8	12	15	19
61	0·5543	·5520	·5498	·5475	·5452	·5430	·5407	·5384	·5362	·5339	4	8	11	15	19
62	0·5317	·5295	·5272	·5250	·5228	·5206	·5184	·5161	·5139	·5117	4	7	11	15	18
63	0·5095	·5073	·5051	·5029	·5008	·4986	·4964	·4942	·4921	·4899	4	7	11	15	18
64	0·4877	·4856	·4834	·4813	·4791	·4770	·4748	·4727	·4706	·4684	4	7	11	14	18
65	0·4663	·4642	·4621	·4599	·4578	·4557	·4536	·4515	·4494	·4473	4	7	11	14	18
66	0·4452	·4431	·4411	·4390	·4369	·4348	·4327	·4307	·4286	·4265	3	7	10	14	17
67	0·4245	·4224	·4204	·4183	·4163	·4142	·4122	·4101	·4081	·4061	3	7	10	14	17
68	0·4040	·4020	·4000	·3979	·3959	·3939	·3919	·3899	·3879	·3859	3	7	10	13	17
69	0·3839	·3819	·3799	·3779	·3759	·3739	·3719	·3699	·3679	·3659	3	6	10	13	17
70	0·3640	·3620	·3600	·3581	·3561	·3541	·3522	·3502	·3482	·3463	3	6	10	13	16
71	0·3443	·3424	·3404	·3385	·3365	·3346	·3327	·3307	·3288	·3269	3	6	10	13	16
72	0·3249	·3230	·3211	·3191	·3172	·3153	·3134	·3115	·3096	·3076	3	6	9	13	16
73	0·3057	·3038	·3019	·3000	·2981	·2962	·2943	·2924	·2905	·2886	3	6	9	13	16
74	0·2867	·2849	·2830	·2811	·2792	·2773	·2754	·2736	·2717	·2698	3	6	9	13	16
75	0·2679	·2661	·2642	·2623	·2605	·2586	·2568	·2549	·2530	·2512	3	6	9	12	16
76	0·2493	·2475	·2456	·2438	·2419	·2401	·2382	·2364	·2345	·2327	3	6	9	12	15
77	0·2309	·2290	·2272	·2254	·2235	·2217	·2199	·2180	·2162	·2144	3	6	9	12	15
78	0·2126	·2107	·2089	·2071	·2053	·2035	·2016	·1998	·1980	·1962	3	6	9	12	15
79	0·1944	·1926	·1908	·1890	·1871	·1853	·1835	·1817	·1799	·1781	3	6	9	12	15
80	0·1763	·1745	·1727	·1709	·1691	·1673	·1655	·1638	·1620	·1602	3	6	9	12	15
81	0·1584	·1566	·1548	·1530	·1512	·1495	·1477	·1459	·1441	·1423	3	6	9	12	15
82	0·1405	·1388	·1370	·1352	·1334	·1317	·1299	·1281	·1263	·1246	3	6	9	12	15
83	0·1228	·1210	·1192	·1175	·1157	·1139	·1122	·1104	·1086	·1069	3	6	9	12	15
84	0·1051	·1033	·1016	·0998	·0981	·0963	·0945	·0928	·0910	·0892	3	6	9	12	15
85	0·0875	·0857	·0840	·0822	·0805	·0787	·0769	·0752	·0734	·0717	3	6	9	12	15
86	0·0699	·0682	·0664	·0647	·0629	·0612	·0594	·0577	·0559	·0542	3	6	9	12	15
87	0·0524	·0507	·0489	·0472	·0454	·0437	·0419	·0402	·0384	·0367	3	6	9	12	15
88	0·0349	·0332	·0314	·0297	·0279	·0262	·0244	·0227	·0209	·0192	3	6	9	12	15
89	0·0175	·0157	·0140	·0122	·0105	·0087	·0070	·0052	·0035	·0017	3	6	9	12	15
	0′	6′	12′	18′	24′	30′	36′	42′	48′	54′	1′	2′	3′	4′	5′

LOGARITHMIC SINES

	0′	6′	12′	18′	24′	30′	36′	42′	48′	54′	1′	2′	3′	4′	5′
0°	− ∞	3̄·2419	·5429	·7190	·8439	3̄·9408	2̄·0200	·0870	·1450	·1961					
1	2̄·2419	·2832	·3210	·3558	·3880	·4179	·4459	·4723	·4971	·5206	p.p. cease				
2	2̄·5428	·5640	·5842	·6035	·6220	·6397	·6567	·6731	·6889	·7041	to be				
3	2̄·7188	·7330	·7468	·7602	·7731	·7857	·7979	·8098	·8213	·8326	sufficiently				
4	2̄·8436	·8543	·8647	·8749	·8849	·8946	·9042	·9135	·9226	·9315	accurate				
5	2̄·9403	·9489	·9573	·9655	·9736	·9816	·9894	2̄·9970	1̄·0046	·0120	13	26	39	53	66
6	1̄·0192	·0264	·0334	·0403	·0472	·0539	·0605	·0670	·0734	·0797	11	22	33	44	56
7	1̄·0859	·0920	·0981	·1040	·1099	·1157	·1214	·1271	·1326	·1381	10	19	29	38	48
8	1̄·1436	·1489	·1542	·1594	·1646	·1697	·1747	·1797	·1847	·1895	8	17	25	34	42
9	1̄·1943	·1991	·2038	·2085	·2131	·2176	·2221	·2266	·2310	·2353	8	15	23	30	38
10	1̄·2397	·2439	·2482	·2524	·2565	·2606	·2647	·2687	·2727	·2767	7	14	20	27	34
11	1̄·2806	·2845	·2883	·2921	·2959	·2997	·3034	·3070	·3107	·3143	6	12	19	25	31
12	1̄·3179	·3214	·3250	·3284	·3319	·3353	·3387	·3421	·3455	·3488	6	11	17	23	28
13	1̄·3521	·3554	·3586	·3618	·3650	·3682	·3713	·3745	·3775	·3806	5	11	16	21	26
14	1̄·3837	·3867	·3897	·3927	·3957	·3986	·4015	·4044	·4073	·4102	5	10	15	20	24
15	1̄·4130	·4158	·4186	·4214	·4242	·4269	·4296	·4323	·4350	·4377	5	9	14	18	23
16	1̄·4403	·4430	·4456	·4482	·4508	·4533	·4559	·4584	·4609	·4634	4	9	13	17	21
17	1̄·4659	·4684	·4709	·4733	·4757	·4781	·4805	·4829	·4853	·4876	4	8	12	16	20
18	1̄·4900	·4923	·4946	·4969	·4992	·5015	·5037	·5060	·5082	·5104	4	8	11	15	19
19	1̄·5126	·5148	·5170	·5192	·5213	·5235	·5256	·5278	·5299	·5320	4	7	11	14	18
20	1̄·5341	·5361	·5382	·5402	·5423	·5443	·5463	·5484	·5504	·5523	3	7	10	13	17
21	1̄·5543	·5563	·5583	·5602	·5621	·5641	·5660	·5679	·5698	·5717	3	6	10	13	16
22	1̄·5736	·5754	·5773	·5792	·5810	·5828	·5847	·5865	·5883	·5901	3	6	9	12	15
23	1̄·5919	·5937	·5954	·5972	·5990	·6007	·6024	·6042	·6059	·6076	3	6	9	12	14
24	1̄·6093	·6110	·6127	·6144	·6161	·6177	·6194	·6210	·6227	·6243	3	6	8	11	14
25	1̄·6259	·6276	·6292	·6308	·6324	·6340	·6356	·6371	·6387	·6403	3	5	8	11	13
26	1̄·6418	·6434	·6449	·6465	·6480	·6495	·6510	·6526	·6541	·6556	3	5	8	10	13
27	1̄·6570	·6585	·6600	·6615	·6629	·6644	·6659	·6673	·6687	·6702	2	5	7	10	12
28	1̄·6716	·6730	·6744	·6759	·6773	·6787	·6801	·6814	·6828	·6842	2	5	7	9	12
29	1̄·6856	·6869	·6883	·6896	·6910	·6923	·6937	·6950	·6963	·6977	2	4	7	9	11
30	1̄·6990	·7003	·7016	·7029	·7042	·7055	·7068	·7080	·7093	·7106	2	4	6	9	11
31	1̄·7118	·7131	·7144	·7156	·7168	·7181	·7193	·7205	·7218	·7230	2	4	6	8	10
32	1̄·7242	·7254	·7266	·7278	·7290	·7302	·7314	·7326	·7338	·7349	2	4	6	8	10
33	1̄·7361	·7373	·7384	·7396	·7407	·7419	·7430	·7442	·7453	·7464	2	4	6	8	10
34	1̄·7476	·7487	·7498	·7509	·7520	·7531	·7542	·7553	·7564	·7575	2	4	6	7	9
35	1̄·7586	·7597	·7607	·7618	·7629	·7640	·7650	·7661	·7671	·7682	2	4	5	7	9
36	1̄·7692	·7703	·7713	·7723	·7734	·7744	·7754	·7764	·7774	·7785	2	3	5	7	9
37	1̄·7795	·7805	·7815	·7825	·7835	·7844	·7854	·7864	·7874	·7884	2	3	5	7	8
38	1̄·7893	·7903	·7913	·7922	·7932	·7941	·7951	·7960	·7970	·7979	2	3	5	6	8
39	1̄·7989	·7998	·8007	·8017	·8026	·8035	·8044	·8053	·8063	·8072	2	3	5	6	8
40	1̄·8081	·8090	·8099	·8108	·8117	·8125	·8134	·8143	·8152	·8161	1	3	4	6	7
41	1̄·8169	·8178	·8187	·8195	·8204	·8213	·8221	·8230	·8238	·8247	1	3	4	6	7
42	1̄·8255	·8264	·8272	·8280	·8289	·8297	·8305	·8313	·8322	·8330	1	3	4	6	7
43	1̄·8338	·8436	·8354	·8362	·8370	·8378	·8386	·8394	·8402	·8410	1	3	4	5	7
44	1̄·8418	·8426	·8433	·8441	·8449	·8457	·8464	·8472	·8480	·8487	1	3	4	5	6
	0′	6′	12′	18′	24′	30′	36′	42′	48′	54′	1′	2′	3′	4′	5′

LOGARITHMIC SINES

	0′	6′	12′	18′	24′	30′	36′	42′	48′	54′	1′	2′	3′	4′	5′
45°	T̄·8495	·8502	·8510	·8517	·8525	·8532	·8540	·8547	·8555	·8562	1	2	4	5	6
46	T̄·8569	·8577	·8584	·8591	·8598	·8606	·8613	·8620	·8627	·8634	1	2	4	5	6
47	T̄·8641	·8648	·8655	·8662	·8669	·8676	·8683	·8690	·8697	·8704	1	2	4	5	6
48	T̄·8711	·8718	·8724	·8731	·8738	·8745	·8751	·8758	·8765	·8771	1	2	3	4	6
49	T̄·8778	·8784	·8791	·8797	·8804	·8810	·8817	·8823	·8830	·8836	1	2	3	4	5
50	T̄·8843	·8849	·8855	·8862	·8868	·8874	·8880	·8887	·8893	·8899	1	2	3	4	5
51	T̄·8905	·8911	·8917	·8923	·8929	·8935	·8941	·8947	·8953	·8959	1	2	3	4	5
52	T̄·8965	·8971	·8977	·8983	·8989	·8995	·9000	·9006	·9012	·9018	1	2	3	4	5
53	T̄·9023	·9029	·9035	·9041	·9046	·9052	·9057	·9063	·9069	·9074	1	2	3	4	5
54	T̄·9080	·9085	·9091	·9096	·9101	·9107	·9112	·9118	·9123	·9128	1	2	3	4	5
55	T̄·9134	·9139	·9144	·9149	·9155	·9160	·9165	·9170	·9175	·9181	1	2	3	3	4
56	T̄·9186	·9191	·9196	·9201	·9206	·9211	·9216	·9221	·9226	·9231	1	2	3	3	4
57	T̄·9236	·9241	·9246	·9251	·9255	·9260	·9265	·9270	·9275	·9279	1	2	2	3	4
58	T̄·9284	·9289	·9294	·9298	·9303	·9308	·9312	·9317	·9322	·9326	1	2	2	3	4
59	T̄·9331	·9335	·9340	·9344	·9349	·9353	·9358	·9362	·9367	·9371	1	1	2	3	4
60	T̄·9375	·9380	·9384	·9388	·9393	·9397	·9401	·9406	·9410	·9414	1	1	2	3	4
61	T̄·9418	·9422	·9427	·9431	·9435	·9439	·9443	·9447	·9451	·9455	1	1	2	3	3
62	T̄·9459	·9463	·9467	·9471	·9475	·9479	·9483	·9487	·9491	·9495	1	1	2	3	3
63	T̄·9499	·9503	·9506	·9510	·9514	·9518	·9522	·9525	·9529	·9533	1	1	2	3	3
64	T̄·9537	·9540	·9544	·9548	·9551	·9555	·9558	·9562	·9566	·9569	1	1	2	2	3
65	T̄·9573	·9576	·9580	·9583	·9587	·9590	·9594	·9597	·9601	·9604	1	1	2	2	3
66	T̄·9607	·9611	·9614	·9617	·9621	·9624	·9627	·9631	·9634	·9637	1	1	2	2	3
67	T̄·9640	·9643	·9647	·9650	·9653	·9656	·9659	·9662	·9665	·9669	1	1	2	2	3
68	T̄·9672	·9675	·9678	·9681	·9684	·9687	·9690	·9693	·9696	·9699	1	1	2	2	2
69	T̄·9702	·9704	·9707	·9710	·9713	·9716	·9719	·9722	·9724	·9727	0	1	1	2	2
70	T̄·9730	·9733	·9735	·9738	·9741	·9743	·9746	·9749	·9751	·9754	0	1	1	2	2
71	T̄·9757	·9759	·9762	·9764	·9767	·9770	·9772	·9775	·9777	·9780	0	1	1	2	2
72	T̄·9782	·9785	·9787	·9789	·9792	·9794	·9797	·9799	·9801	·9804	0	1	1	2	2
73	T̄·9806	·9808	·9811	·9813	·9815	·9817	·9820	·9822	·9824	·9826	0	1	1	1	2
74	T̄·9828	·9831	·9833	·9835	·9837	·9839	·9841	·9843	·9845	·9847	0	1	1	1	2
75	T̄·9849	·9851	·9853	·9855	·9857	·9859	·9861	·9863	·9865	·9867	0	1	1	1	2
76	T̄·9869	·9871	·9873	·9875	·9876	·9878	·9880	·9882	·9884	·9885	0	1	1	1	2
77	T̄·9887	·9889	·9891	·9892	·9894	·9896	·9897	·9899	·9901	·9902	0	1	1	1	1
78	T̄·9904	·9906	·9907	·9909	·9910	·9912	·9913	·9915	·9916	·9918	0	1	1	1	1
79	T̄·9919	·9921	·9922	·9924	·9925	·9927	·9928	·9929	·9931	·9932	0	0	1	1	1
80	T̄·9934	·9935	·9936	·9937	·9939	·9940	·9941	·9943	·9944	·9945	0	0	1	1	1
81	T̄·9946	·9947	·9949	·9950	·9951	·9952	·9953	·9954	·9955	·9956	0	0	1	1	1
82	T̄·9958	·9959	·9960	·9961	·9962	·9963	·9964	·9965	·9966	·9967	0	0	0	1	1
83	T̄·9968	·9968	·9969	·9970	·9971	·9972	·9973	·9974	·9975	·9975	0	0	0	1	1
84	T̄·9976	·9977	·9978	·9978	·9979	·9980	·9981	·9981	·9982	·9983	0	0	0	0	1
85	T̄·9983	·9984	·9985	·9985	·9986	·9987	·9987	·9988	·9988	·9989	0	0	0	0	0
86	T̄·9989	·9990	·9990	·9991	·9991	·9992	·9992	·9993	·9993	·9994	0	0	0	0	0
87	T̄·9994	·9994	·9995	·9995	·9996	·9996	·9996	·9996	·9997	·9997	0	0	0	0	0
88	T̄·9997	·9998	·9998	·9998	·9999	·9999	·9999	·9999	·9999	·9999	0	0	0	0	0
89	T̄·9999	T̄·9999	0·0000	·0000	·0000	·0000	·0000	·0000	·0000	·0000	0	0	0	0	0
	0′	6′	12′	18′	24′	30′	36′	42′	48′	54′	1′	2′	3′	4′	5′

Proportional Parts Subtract

	0′	6′	12′	18′	24′	30′	36′	42′	48′	54′	1′	2′	3′	4′	5′
0°	0·0000	·0000	·0000	·0000	·0000	·0000	·0000	·0000	0·0000	T·9999	0	0	0	0	0
1	T·9999	·9999	·9999	·9999	·9999	·9999	·9998	·9998	·9998	·9998	0	0	0	0	0
2	T·9997	·9997	·9997	·9996	·9996	·9996	·9996	·9995	·9995	·9994	0	0	0	0	0
3	T·9994	·9994	·9993	·9993	·9992	·9992	·9991	·9991	·9990	·9990	0	0	0	0	0
4	T·9989	·9989	·9988	·9988	·9987	·9987	·9986	·9985	·9985	·9984	0	0	0	0	0
5	T·9983	·9983	·9982	·9981	·9981	9980	·9979	·9978	·9978	·9977	0	0	0	0	1
6	T·9976	·9975	·9975	·9974	·9973	·9972	·9971	·9970	·9969	·9968	0	0	0	1	1
7	T·9968	·9967	·9966	·9965	·9964	·9963	·9962	·9961	·9960	·9959	0	0	0	1	1
8	T·9958	·9956	·9955	·9954	·9953	·9952	·9951	·9950	·9949	·9947	0	0	1	1	1
9	T·9946	·9945	·9944	·9943	·9941	·9940	·9939	·9937	·9936	·9935	0	0	1	1	1
10	T·9934	·9932	·9931	·9929	·9928	·9927	·9925	·9924	·9922	·9921	0	0	1	1	1
11	T·9919	·9918	·9916	·9915	·9913	·9912	·9910	·9909	·9907	·9906	0	1	1	1	1
12	T·9904	·9902	·9901	·9899	·9897	·9896	·9894	·9892	·9891	·9889	0	1	1	1	1
13	T·9887	·9885	·9884	·9882	·9880	·9878	·9876	·9875	·9873	·9871	0	1	1	1	2
14	T·9869	·9867	·9865	·9863	·9861	·9859	·9857	·9855	·9853	·9851	0	1	1	1	2
15	T·9849	·9847	·9845	·9843	·9841	·9839	·9837	·9835	·9833	·9831	0	1	1	1	2
16	T·9828	·9826	·9824	·9822	·9820	·9817	·9815	·9813	·9811	·9808	0	1	1	1	2
17	T·9806	·9804	·9801	·9799	·9797	·9794	·9792	·9789	·9787	·9785	0	1	1	2	2
18	T·9782	·9780	·9777	·9775	·9772	·9770	·9767	·9764	·9762	·9759	0	1	1	2	2
19	T·9757	·9754	·9751	·9749	·9746	·9743	·9741	·9738	·9735	·9733	0	1	1	2	2
20	T·9730	·9727	·9724	·9722	·9719	·9716	·9713	·9710	·9707	·9704	0	1	1	2	2
21	T·9702	·9699	·9696	·9693	·9690	·9687	·9684	·9681	·9678	·9675	1	1	2	2	2
22	T·9672	·9669	·9665	·9662	·9659	·9656	·9653	·9650	·9647	·9643	1	1	2	2	3
23	T·9640	·9637	·9634	·9631	·9627	·9624	·9621	·9617	·9614	·9611	1	1	2	2	3
24	T·9607	·9604	·9601	·9597	·9594	·9590	·9587	·9583	·9580	·9576	1	1	2	2	3
25	T·9573	·9569	·9566	·9562	·9558	·9555	·9551	·9548	·9544	·9540	1	1	2	2	3
26	T·9537	·9533	·9529	·9525	·9522	·9518	·9514	·9510	·9506	·9503	1	1	2	3	3
27	T·9499	·9495	·9491	·9487	·9483	·9479	·9475	·9471	·9467	·9463	1	1	2	3	3
28	T·9459	·9455	·9451	·9447	·9443	·9439	·9435	·9431	·9427	·9422	1	1	2	3	3
29	T·9418	·9414	·9410	·9406	·9401	·9397	·9393	·9388	·9384	·9380	1	1	2	3	4
30	T·9375	·9371	·9367	·9362	·9358	·9353	·9349	·9344	·9340	·9335	1	1	2	3	4
31	T·9331	·9326	·9322	·9317	·9312	·9308	·9303	·9298	·9294	·9289	1	2	2	3	4
32	T·9284	·9279	·9275	·9270	·9265	·9260	·9255	·9251	·9246	·9241	1	2	2	3	4
33	T·9236	·9231	·9226	·9221	·9216	·9211	·9206	·9201	·9196	·9191	1	2	3	3	4
34	T·9186	·9181	·9175	·9170	·9165	·9160	·9155	·9149	·9144	·9139	1	2	3	3	4
35	T·9134	·9128	·9123	·9118	·9112	·9107	·9101	·9096	·9091	·9085	1	2	3	4	5
36	T·9080	·9074	·9069	·9063	·9057	·9052	·9046	·9041	·9035	·9029	1	2	3	4	5
37	T·9023	·9018	·9012	·9006	·9000	·8995	·8989	·8983	·8977	·8971	1	2	3	4	5
38	T·8965	·8959	·8953	·8947	·8941	·8935	·8929	·8923	·8917	·8911	1	2	3	4	5
39	T·8905	·8899	·8893	·8887	·8880	·8874	·8868	·8862	·8855	·8849	1	2	3	4	5
40	T·8843	·8836	·8830	·8823	·8817	·8810	·8804	·8797	·8791	·8784	1	2	3	4	5
41	T·8778	·8771	·8765	·8758	·8751	·8745	·8738	·8731	·8724	·8718	1	2	3	4	6
42	T·8711	·8704	·8697	·8690	·8683	·8676	·8669	·8662	·8655	·8648	1	2	4	5	6
43	T·8641	·8634	·8627	·8620	·8613	·8606	·8598	·8591	·8584	·8577	1	2	4	5	6
44	T·8569	·8562	·8555	·8547	·8540	·8532	·8525	·8517	·8510	·8502	1	2	4	5	6
	0′	6′	12′	18′	24′	30′	36′	42′	48′	54′	1′	2′	3′	4′	5′

LOGARITHMIC COSINES

Proportional Parts Subtract

	0′	6′	12′	18′	24′	30′	36′	42′	48′	54′	1′	2′	3′	4′	5′
45°	1̄·8495	·8487	·8480	·8472	·8464	·8457	·8449	·8441	·8433	·8426	1	3	4	5	6
46	1̄·8418	·8410	·8402	·8394	·8386	·8378	·8370	·8362	·8354	·8346	1	3	4	5	7
47	1̄·8338	·8330	·8322	·8313	·8305	·8297	·8289	·8280	·8272	·8264	1	3	4	6	7
48	1̄·8255	·8247	·8238	·8230	·8221	·8213	·8204	·8195	·8187	·8178	1	3	4	6	7
49	1̄·8169	·8161	·8152	·8143	·8134	·8125	·8117	·8108	·8099	·8090	1	3	4	6	7
50	1̄·8081	·8072	·8063	·8053	·8044	·8035	·8026	·8017	·8007	·7998	2	3	5	6	8
51	1̄·7989	·7979	·7970	·7960	·7951	·7941	·7932	·7922	·7913	·7903	2	3	5	6	8
52	1̄·7893	·7884	·7874	·7864	·7854	·7844	·7835	·7825	·7815	·7805	2	3	5	7	8
53	1̄·7795	·7785	·7774	·7764	·7754	·7744	·7734	·7723	·7713	·7703	2	3	5	7	9
54	1̄·7692	·7682	·7671	·7661	·7650	·7640	·7629	·7618	·7607	·7597	2	4	5	7	9
55	1̄·7586	·7575	·7564	·7553	·7542	·7531	·7520	·7509	·7498	·7487	2	4	6	7	9
56	1̄·7476	·7464	·7453	·7442	·7430	·7419	·7407	·7396	·7384	·7373	2	4	6	8	10
57	1̄·7361	·7349	·7338	·7326	·7314	·7302	·7290	·7278	·7266	·7254	2	4	6	8	10
58	1̄·7242	·7230	·7218	·7205	·7193	·7181	·7168	·7156	·7144	·7131	2	4	6	8	10
59	1̄·7118	·7106	·7093	·7080	·7068	·7055	·7042	·7029	·7016	·7003	2	4	6	9	11
60	1̄·6990	·6977	·6963	·6950	·6937	·6923	·6910	·6896	·6883	·6869	2	4	7	9	11
61	1̄·6856	·6842	·6828	·6814	·6801	·6787	·6773	·6759	·6744	·6730	2	5	7	9	12
62	1̄·6716	·6702	·6687	·6673	·6659	·6644	·6629	·6615	·6600	·6585	2	5	7	10	12
63	1̄·6570	·6556	·6541	·6526	·6510	·6495	·6480	·6465	·6449	·6434	3	5	8	10	13
64	1̄·6418	·6403	·6387	·6371	·6356	·6340	·6324	·6308	·6292	·6276	3	5	8	11	13
65	1̄·6259	·6243	·6227	·6210	·6194	·6177	·6161	·6144	·6127	·6110	3	6	8	11	14
66	1̄·6093	·6076	·6059	·6042	·6024	·6007	·5990	·5972	·5954	·5937	3	6	9	12	14
67	1̄·5919	·5901	·5883	·5865	·5847	·5828	·5810	·5792	·5773	·5754	3	6	9	12	15
68	1̄·5736	·5717	·5698	·5679	·5660	·5641	·5621	·5602	·5583	·5563	3	6	10	13	16
69	1̄·5543	·5523	·5504	·5484	·5463	·5443	·5423	·5402	·5382	·5361	3	7	10	13	17
70	1̄·5341	·5320	·5299	·5278	·5256	·5235	·5213	·5192	·5170	·5148	4	7	11	14	18
71	1̄·5126	·5104	·5082	·5060	·5037	·5015	·4992	·4969	·4946	·4923	4	8	11	15	19
72	1̄·4900	·4876	·4853	·4829	·4805	·4781	·4757	·4733	·4709	·4684	4	8	12	16	20
73	1̄·4659	·4634	·4609	·4584	·4559	·4533	·4508	·4482	·4456	·4430	4	9	13	17	21
74	1̄·4403	·4377	·4350	·4323	·4296	·4269	·4242	·4214	·4186	·4158	5	9	14	18	23
75	1̄·4130	·4102	·4073	·4044	·4015	·3986	·3957	·3927	·3897	·3867	5	10	15	20	24
76	1̄·3837	·3806	·3775	·3745	·3713	·3682	·3650	·3618	·3586	·3554	5	11	16	21	26
77	1̄·3521	·3488	·3455	·3421	·3387	·3353	·3319	·3284	·3250	·3214	6	11	17	23	28
78	1̄·3179	·3143	·3107	·3070	·3034	·2997	·2959	·2921	·2883	·2845	6	12	19	25	31
79	1̄·2806	·2767	·2727	·2687	·2647	·2606	·2565	·2524	·2482	·2439	7	14	20	27	34
80	1̄·2397	·2353	·2310	·2266	·2221	·2176	·2131	·2085	·2038	·1991	8	15	23	30	38
81	1̄·1943	·1895	·1847	·1797	·1747	·1697	·1646	·1594	·1542	·1489	8	17	25	34	42
82	1̄·1436	·1381	·1326	·1271	·1214	·1157	·1099	·1040	·0981	·0920	10	19	29	38	48
83	1̄·0859	·0797	·0734	·0670	·0605	·0539	·0472	·0403	·0334	·0264	11	22	33	44	56
84	1̄·0192	·0120	1̄·0046	2̄·9970	·9894	·9816	·9736	·9655	·9573	·9489	13	26	39	53	66
85	2̄·9403	·9315	·9226	·9135	·9042	·8946	·8849	·8749	·8647	·8543	p.p. cease				
86	2̄·8436	·8326	·8213	·8098	·7979	·7857	·7731	·7602	·7468	·7330	to be				
87	2̄·7188	·7041	·6889	·6731	·6567	·6397	·6220	·6035	·5842	·5640	sufficiently				
88	2̄·5428	·5206	·4971	·4723	·4459	·4179	·3880	·3558	·3210	·2832	accurate				
89	2̄·2419	·1961	·1450	·0870	2̄·0200	3̄·9408	·8439	·7190	·5429	·2419					
	0′	6′	12′	18′	24′	30′	36′	42′	48′	54′	1′	2′	3′	4′	5′

LOGARITHMIC TANGENTS

Proportional Parts

	0′	6′	12′	18′	24′	30′	36′	42′	48′	54′	1′	2′	3′	4′	5′
0°	− ∞	3.2419	.5429	.7190	.8439	3.9409	2.0200	.0870	.1450	.1962					
1	2.2419	.2833	.3211	.3559	.3881	.4181	.4461	.4725	.4973	.5208		p.p. cease			
2	2.5431	.5643	.5845	.6038	.6223	.6401	.6571	.6736	.6894	.7046		to be			
3	2.7194	.7337	.7475	.7609	.7739	.7865	.7988	.8107	.8223	.8336		sufficiently			
4	2.8446	.8554	.8659	.8762	.8862	.8960	.9056	.9150	.9241	.9331		accurate			
5	2.9420	.9506	.9591	.9674	.9756	.9836	.9915	2.9992	1.0068	.0143	13	27	40	53	66
6	1.0216	.0289	.0360	.0430	.0499	.0567	.0633	.0699	.0764	.0828	11	22	34	45	56
7	1.0891	.0954	.1015	.1076	.1135	.1194	.1252	.1310	.1367	.1423	10	20	29	39	49
8	1.1478	.1533	.1587	.1640	.1693	.1745	.1797	.1848	.1898	.1948	9	17	26	35	43
9	1.1997	.2046	.2094	.2142	.2189	.2236	.2282	.2328	.2374	.2419	8	16	23	31	39
10	1.2463	.2507	.2551	.2594	.2637	.2680	.2722	.2764	.2805	.2846	7	14	21	28	35
11	1.2887	.2927	.2967	.3006	.3046	.3085	.3123	.3162	.3200	.3237	6	13	19	26	32
12	1.3275	.3312	.3349	.3385	.3422	.3458	.3493	.3529	.3564	.3599	6	12	18	24	30
13	1.3634	.3668	.3702	.3736	.3770	.3804	.3837	.3870	.3903	.3935	6	11	17	22	28
14	1.3968	.4000	.4032	.4064	.4095	.4127	.4158	.4189	.4220	.4250	5	10	16	21	26
15	1.4281	.4311	.4341	.4371	.4400	.4430	.4459	.4488	.4517	.4546	5	10	15	20	24
16	1.4575	.4603	.4632	.4660	.4688	.4716	.4744	.4771	.4799	.4826	5	9	14	19	23
17	1.4853	.4880	.4907	.4934	.4961	.4987	.5014	.5040	.5066	.5092	4	9	13	18	22
18	1.5118	.5143	.5169	.5195	.5220	.5245	.5270	.5295	.5320	.5345	4	8	13	17	21
19	1.5370	.5394	.5419	.5443	.5467	.5491	.5516	.5539	.5563	.5587	4	8	12	16	20
20	1.5611	.5634	.5658	.5681	.5704	.5727	.5750	.5773	.5796	.5819	4	8	12	15	19
21	1.5842	.5864	.5887	.5909	.5932	.5954	.5976	.5998	.6020	.6042	4	7	11	15	18
22	1.6064	.6086	.6108	.6129	.6151	.6172	.6194	.6215	.6236	.6257	4	7	11	14	18
23	1.6279	.6300	.6321	.6341	.6362	.6383	.6404	.6424	.6445	.6465	3	7	10	14	17
24	1.6486	.6506	.6527	.6547	.6567	.6587	.6607	.6627	.6647	.6667	3	7	10	13	17
25	1.6687	.6706	.6726	.6746	.6765	.6785	.6804	.6824	.6843	.6863	3	6	10	13	16
26	1.6882	.6901	.6920	.6939	.6958	.6977	.6996	.7015	.7034	.7053	3	6	10	13	16
27	1.7072	.7090	.7109	.7128	.7146	.7165	.7183	.7202	.7220	.7238	3	6	9	12	15
28	1.7257	.7275	.7293	.7311	.7330	.7348	.7366	.7384	.7402	.7420	3	6	9	12	15
29	1.7438	.7455	.7473	.7491	.7509	.7526	.7544	.7562	.7579	.7597	3	6	9	12	15
30	1.7614	.7632	.7649	.7667	.7684	.7701	.7719	.7736	.7753	.7771	3	6	9	12	15
31	1.7788	.7805	.7822	.7839	.7856	.7873	.7890	.7907	.7924	.7941	3	6	9	11	14
32	1.7958	.7975	.7992	.8008	.8025	.8042	.8059	.8075	.8092	.8109	3	6	8	11	14
33	1.8125	.8142	.8158	.8175	.8191	.8208	.8224	.8241	.8257	.8274	3	6	8	11	14
34	1.8290	.8306	.8323	.8339	.8355	.8371	.8388	.8404	.8420	.8436	3	5	8	11	13
35	1.8452	.8468	.8484	.8501	.8517	.8533	.8549	.8565	.8581	.8597	3	5	8	11	13
36	1.8613	.8629	.8644	.8660	.8676	.8692	.8708	.8724	.8740	.8755	3	5	8	11	13
37	1.8771	.8787	.8803	.8818	.8834	.8850	.8865	.8881	.8897	.8912	3	5	8	10	13
38	1.8928	.8944	.8959	.8975	.8990	.9006	.9022	.9037	.9053	.9068	3	5	8	10	13
39	1.9084	.9099	.9115	.9130	.9146	.9161	.9176	.9192	.9207	.9223	3	5	8	10	13
40	1.9238	.9254	.9269	.9284	.9300	.9315	.9330	.9346	.9361	.9376	3	5	8	10	13
41	1.9392	.9407	.9422	.9438	.9453	.9468	.9483	.9499	.9514	.9529	3	5	8	10	13
42	1.9544	.9560	.9575	.9590	.9605	.9621	.9636	.9651	.9666	.9681	3	5	8	10	13
43	1.9697	.9712	.9727	.9742	.9757	.9772	.9788	.9803	.9818	.9833	3	5	8	10	13
44	1.9848	.9864	.9879	.9894	.9909	.9924	.9939	.9955	.9970	.9985	3	5	8	10	13
	0′	6′	12′	18′	24′	30′	36′	42′	48′	54′	1′	2′	3′	4′	5′

LOGARITHMIC TANGENTS

	0′	6′	12′	18′	24′	30′	36′	42′	48′	54′	1′	2′	3′	4′	5′
45°	0·0000	·0015	·0030	·0045	·0061	·0076	·0091	·0106	·0121	·0136	3	5	8	10	13
46	0·0152	·0167	·0182	·0197	·0212	·0228	·0243	·0258	·0273	·0288	3	5	8	10	13
47	0·0303	·0319	·0334	·0349	·0364	·0379	·0395	·0410	·0425	·0440	3	5	8	10	13
48	0·0456	·0471	·0486	·0501	·0517	·0532	·0547	·0562	·0578	·0593	3	5	8	10	13
49	0·0608	·0624	·0639	·0654	·0670	·0685	·0700	·0716	·0731	·0746	3	5	8	10	13
50	0·0762	·0777	·0793	·0808	·0824	·0839	·0854	·0870	·0885	·0901	3	5	8	10	13
51	0·0916	·0932	·0947	·0963	·0978	·0994	·1010	·1025	·1041	·1056	3	5	8	10	13
52	0·1072	·1088	·1103	·1119	·1135	·1150	·1166	·1182	·1197	·1213	3	5	8	10	13
53	0·1229	·1245	·1260	·1276	·1292	·1308	·1324	·1340	·1356	·1371	3	5	8	11	13
54	0·1387	·1403	·1419	·1435	·1451	·1467	·1483	·1499	·1516	·1532	3	5	8	11	13
55	0·1548	·1564	·1580	·1596	·1612	·1629	·1645	·1661	·1677	·1694	3	5	8	11	13
56	0·1710	·1726	·1743	·1759	·1776	·1792	·1809	·1825	·1842	·1858	3	6	8	11	14
57	0·1875	·1891	·1908	·1925	·1941	·1958	·1975	·1992	·2008	·2025	3	6	8	11	14
58	0·2042	·2059	·2076	·2093	·2110	·2127	·2144	·2161	·2178	·2195	3	6	9	11	14
59	0·2212	·2229	·2247	·2264	·2281	·2299	·2316	·2333	·2351	·2368	3	6	9	12	15
60	0·2386	·2403	·2421	·2438	·2456	·2474	·2491	·2509	·2527	·2545	3	6	9	12	15
61	0·2562	·2580	·2598	·2616	·2634	·2652	·2670	·2689	·2707	·2725	3	6	9	12	15
62	0·2743	·2762	·2780	·2798	·2817	·2835	·2854	·2872	·2891	·2910	3	6	9	12	15
63	0·2928	·2947	·2966	·2985	·3004	·3023	·3042	·3061	·3080	·3099	3	6	10	13	16
64	0·3118	·3137	·3157	·3176	·3196	·3215	·3235	·3254	·3274	·3294	3	6	10	13	16
65	0·3313	·3333	·3353	·3373	·3393	·3413	·3433	·3453	·3473	·3494	3	7	10	13	17
66	0·3514	·3535	·3555	·3576	·3596	·3617	·3638	·3659	·3679	·3700	3	7	10	14	17
67	0·3721	·3743	·3764	·3785	·3806	·3828	·3849	·3871	·3892	·3914	4	7	11	14	18
68	0·3936	·3958	·3980	·4002	·4024	·4046	·4068	·4091	·4113	·4136	4	7	11	15	18
69	0·4158	·4181	·4202	·4227	·4250	·4273	·4296	·4319	·4342	·4366	4	8	12	15	19
70	0·4389	·4413	·4437	·4461	·4484	·4509	·4533	·4557	·4581	·4606	4	8	12	16	20
71	0·4630	·4655	·4680	·4705	·4730	·4755	·4780	·4805	·4831	·4857	4	8	13	17	21
72	0·4882	·4908	·4934	·4960	·4986	·5013	·5039	·5066	·5093	·5120	4	9	13	18	22
73	0·5147	·5174	·5201	·5229	·5256	·5284	·5312	·5340	·5368	·5397	5	9	14	19	23
74	0·5425	·5454	·5483	·5512	·5541	·5570	·5600	·5629	·5659	·5689	5	10	15	20	24
75	0·5719	·5750	·5780	·5811	·5842	·5873	·5905	·5936	·5968	·6000	5	10	16	21	26
76	0·6032	·6065	·6097	·6130	·6163	·6196	·6230	·6264	·6298	·6332	6	11	17	22	28
77	0·6366	·6401	·6436	·6471	·6507	·6542	·6578	·6615	·6651	·6688	6	12	18	24	30
78	0·6725	·6763	·6800	·6838	·6877	·6915	·6954	·6994	·7033	·7073	6	13	19	26	32
79	0·7113	·7154	·7195	·7236	·7278	·7320	·7363	·7406	·7449	·7493	7	14	21	28	35
80	0·7537	·7581	·7626	·7672	·7718	·7764	·7811	·7858	·7906	·7954	8	16	23	31	39
81	0·8003	·8052	·8102	·8152	·8203	·8255	·8307	·8360	·8413	·8467	9	17	26	35	43
82	0·8522	·8577	·8633	·8690	·8748	·8806	·8865	·8924	·8985	·9046	10	20	29	39	49
83	0·9109	·9172	·9236	·9301	·9367	·9433	·9501	·9570	·9640	·9711	11	22	34	45	56
84	0·9784	·9857	0·9932	1·0008	·0085	·0164	·0244	·0326	·0409	·0494	13	27	40	53	66
85	1·0580	·0669	·0759	·0850	·0944	·1040	·1138	·1238	·1341	·1446					
86	1·1554	·1664	·1777	·1893	·2012	·2135	·2261	·2391	·2525	·2663					
87	1·2806	·2954	·3106	·3264	·3429	·3599	·3777	·3962	·4155	·4357					
88	1·4569	·4792	·5027	·5275	·5539	·5819	·6119	·6441	·6789	·7167					
89	1·7581	·8038	·8550	·9130	1·9800	2·0591	·1561	·2810	·4571	·7581					

p.p. cease to be sufficiently accurate

| | 0′ | 6′ | 12′ | 18′ | 24′ | 30′ | 36′ | 42′ | 48′ | 54′ | 1′ | 2′ | 3′ | 4′ | 5′ |

ANSWERS

p. 39   Exercise 1

1. (1) a^{10} (3) x^{12} (5) $2^7 = 128$ (6) $3^7 = 2187$
   (2) b^{12} (4) $\frac{3}{32}x^{10}$
2. (1) a^4 (2) c^5 (3) x^{12} (4) $2^6 = 64$
3. (1) x^6 (2) a^3 (3) $\frac{1}{a}$ (4) x^7
4. (1) a^{14} (4) $2^8 = 256$ (7) $\frac{1}{32}x^{20}$
   (2) x^{12} (5) $10^6 = 1,000,000$ (8) $3^9 = 19683$
   (3) $16b^{16}$ (6) $27a^6$

p. 42   Exercise 2

1. $\sqrt{3}$, $\frac{1}{4}$, $\frac{3}{a^2}$, 1, $\frac{1}{\sqrt{2}}$, 3, $3a^2$, $\sqrt{64}$, $\frac{1}{10^3} = 0{\cdot}001$
2. (1) 5·656   (3) $\frac{1}{10}$ (5) 2·828   (6) 316·2
   (2) 27   (4) $a^{1\frac{1}{2}}$
3. (1) 4   (3) 1000   (5) 16   (6) 31·62
   (2) 125   (4) $\frac{1}{5^6} = \frac{1}{15625}$
4. (1) 4   (2) $2\frac{7}{8}$ (3) 4   (4) $\frac{1}{8}$ (5) 8   (6) $\frac{1}{32}$
5. 5·656
6. (1) $\sqrt[6]{a^5}$ (2) $100\sqrt{10}$

p. 47   Exercise 3

1. 1, 3, 4, 2, 0, 5, 1, 3, 0, 2
2. (1) 0·6990, 1·6990, 2·6990, 4·6990
   (2) 0·6721, 2·6721, 4·6721
   (3) 1·7226, 0·7226, 2·7226
   (4) 2·9768, 0·9768, 4·9768
   (5) 0·7588, 1·9842, 3·8433
3. (1) 446·7, 44670, 44·67   (3) 4·714, 471·4, 471,400
   (2) 87·70, 8770, 8·770   (4) 2628, 5·229, 114·0

p. 49   Exercise 4

1. 344·6   9. 14·22   17. 1·656
2. 276·4   10. 13·56   18. 1436
3. 1397   11. 851·3   19. 1·359
4. 5976   12. 2650   20. 1·695
5. 2·396   13. 3·137   21. 2·321
6. 6·997   14. 728·8   22. 2·786
7. 1·589   15. 2·172   23. 5·002
8. 222·8   16. 104·6   24. 1·546

p. 51   Exercise 5

1. (1) 0·4470, $\bar{1}$·4470, $\bar{2}$·4470   (3) $\bar{3}$·9904, $\bar{4}$·9904, $\bar{1}$·9904
   (2) 0·6298, $\bar{1}$·6298, $\bar{3}$·6298   (4) $\bar{2}$·8097, $\bar{1}$·8097, $\bar{4}$·8097
2. (1) $\bar{2}$·7771   (3) $\bar{3}$·9011   (5) $\bar{1}$·7538
   (2) $\bar{4}$·6749   (4) $\bar{5}$·9673   (6) $\bar{2}$·9023
3. (1) 0·2159   (3) 0·03070   (5) 0·5940
   (2) 0·007453   (4) 0·0004402   (6) $2·482 \times 10^{-8}$

p. 53   Exercise 6

1. (1) $\bar{4}$·6037   (2) $\bar{2}$·7126.
2. (1) $\bar{2}$·5926   (3) $\bar{1}$·6597
   (2) 0·8263   (4) 2·4814
3. (1) $\bar{1}$·7464   (3) $\bar{3}$·8910   (5) $\bar{1}$·1958
   (2) $\bar{4}$·8368   (4) $\bar{1}$·3673   (6) $\bar{4}$·7913
4. (1) $\bar{2}$·6856   (3) $\bar{1}$·7754   (5) $\bar{2}$·0254
   (2) $\bar{1}$·07155   (4) $\bar{1}$·1463   (6) 0·5619
5. (1) $\bar{1}$·7399   (3) $\bar{2}$·7726   (5) $\bar{1}$·7266
   (2) $\bar{1}$·7127   (4) $\bar{2}$·5598   (6) $\bar{3}$·8973
6. 15·42   13. 0·1600   20. 1·457
7. 0·3285   14. 85·23   21. 3·558
8. 0·01529   15. 0·8414   22. 5·471
9. 5·699   16. 0·1226   23. 0·1014
10. 0·6116   17. 1·197   24. 0·1429
11. 0·03239   18. 0·07115   25. 9·399
12. 0·04903   19. 1·826

p. 62   Exercise 7

1. $\tan ABC = \dfrac{AC}{CB} = \dfrac{CD}{DB} = \dfrac{CQ}{QD} = \dfrac{DQ}{QB} = \dfrac{AD}{CD}$

   $\tan CAB = \dfrac{CB}{AC} = \dfrac{DB}{CD} = \dfrac{QD}{CQ} = \dfrac{QB}{DQ} = \dfrac{CD}{AD}$

2. $\tan ABC = \frac{1}{3}$, $\tan CAB = \frac{3}{4}$
3. (1) 0·3249   (3) 1·4826   (5) 0·2549
   (2) 0·9325   (4) 3·2709   (6) 0·6950
4. (1) 0·1635   (3) 0·8122   (5) 2·1123
   (2) 0·6188   (4) 1·3009
5. (1) 28° 36′   (3) 70° 30′   (5) 33° 51′
   (2) 61° 18′   (4) 52° 26′   (6) 14° 16′
6. 8·36 m   7. 67° 23′, 67° 23′, 45° 14′   8. 19·54 m
9. 1·41 km   10. 21·3 m approx.   11. 37°; 53° approx.
12. 144 m

p. 69   Exercise 8

1. $\sin ABC = \dfrac{AC}{AB} = \dfrac{DQ}{DB} = \dfrac{CD}{CB} = \dfrac{CQ}{CD} = \dfrac{AD}{AC}$

   $\sin CAB = \dfrac{CB}{AB} = \dfrac{QB}{DB} = \dfrac{DB}{CB} = \dfrac{DQ}{CD} = \dfrac{CD}{AC}$

   $\cos ABC = \dfrac{CB}{AB} = \dfrac{QB}{DB} = \dfrac{DB}{CB} = \dfrac{DQ}{CD} = \dfrac{CD}{AC}$

   $\cos CAB = \dfrac{AC}{AB} = \dfrac{DQ}{DB} = \dfrac{CD}{CB} = \dfrac{CQ}{CD} = \dfrac{AD}{AC}$

2. Cosine is 0·1109, sine is 0·9939
3. Length is 5·14 cm approx., distance from centre 3·06 cm approx.
4. Sines 0·6 and 0·8, cosines 0·8 and 0·6
5. (1) 0·2521   (2) 0·7400   (3) 0·9353
6. (1) 29° 48′   (2) 30° 46′   (3) 52° 14′
7. (1) 0·9350   (3) 0·4594   (5) 0·1863
   (2) 0·7149   (4) 0·7789   (6) 0·5390
8. (1) 57° 47′   (3) 69° 14′   (5) 37° 43′
   (2) 20° 39′   (4) 77° 27′   (6) 59° 4′
9. 10° 5′   11. 13° 56′
10. 7·34 m; 37° 48′; 52° 12′   12. 47° 36′; 43·8 m approx.

p. 77   Exercise 9

1. (1) 1·7263   (3) 1·3589   (5) 1·2045
   (2) 1·1576   (4) 1·6649   (6) 0·3528
2. (1) 60° 37′   (2) 64° 45′   (3) 69° 18′
3. 48·2 mm   10. (a) 1·869
4. 22° 37′, 67° 23′   (b) 1·56 approx.
5. 2·87 m   11. 0·5602
6. 7·19 m   12. (1) 0·2616
7. (a) 0·3465   (2) − 0·4695
   (b) 0·4394   13. 37° 8′
8. (a) 0·2204   14. 1·2234
   (b) 2·988   15. 0·09661
9. (a) 0·7357   16. 553·5
   (b) 1·691

p. 83   Exercise 10

1. 35° 1′, 54° 59′, 2·86 m   2. 44° 12′
3. $a = 55·5$, $b = 72·6$
4. $A = 30° 30′$, $B = 59° 30′$
5. $AD = 2·66$ cm, $BD = 1·87$ cm, $DC = 2·81$ cm, $AC = 3·87$ cm
6. $A = 44° 8′$, $b = 390$ mm (approx.)
7. 69° 31′, 60°
8. 10·3 km N., 14·7 km E.
9. 0·68 cm   10. $\dfrac{x\sqrt{3}}{2}$; $\dfrac{\sqrt{3}}{2}$, $\frac{1}{2}$
11. 2·60 cm; 2·34 cm (both approx.)
12. 3° 36′   13. 10·2 km W., 11·7 km N
14. 31° 50′ W. of N; 17·1 km

p. 86   Exercise 11

1. 0·7002   2. $\frac{4}{5}$; $\frac{3}{4}$
3. 0·8827   4. 1·6243
5. 0·6745, 0·8290, 0·5592   6. 1·1547
7. 1·9121; 0·5230; 0·8523

8. $\sec \theta = \sqrt{1 + t^2}$; $\cos \theta = \dfrac{1}{\sqrt{1 + t^2}}$; $\sin \theta = \dfrac{t}{\sqrt{1 + t^2}}$

9. $\sin \alpha = 0·8829$; $\tan \alpha = 1·8807$

p. 96   Exercise 12

1. sines are   (a) 0·9781   (c) 0·9428   (e) 0·4289
                      (b) 0·5068   (d) 0·5698
   cosines are   (a) −0·2079   (c) −0·3333   (e) −0·9033
                      (b) −0·8621   (d) −0·8218
   tangents are   (a) −4·7046   (c) −2·8291   (e) −0·4748
                      (b) −0·5879   (d) −0·6933
2. (a) 40° 36′ or 139° 24′.   (c) 20° 18′ or 159° 42′
   (b) 65° 52′ or 114° 8′.   (d) 45° 25′ or 134° 35′
3. (a) 117°   (c) 100° 18′.   (e) 142° 21′
   (b) 144° 24′   (d) 159° 18′.   (f) 156° 15′
4. (a) 151°   (c) 112° 18′.   (e) 144° 28′
   (b) 123° 48′   (d) 119° 36′.   (f) 130° 23′
5. (a) 2·2812   (b) −1·0485.   (c) −3·3122
6. (a) 127° 16′   (d) 24° or 156°
   (b) 118°   (e) 149°
   (c) 35° 18′ or 144° 42′   (f) 110° 54′
7. 0·5530
8. (a) 69° or 111°   (c) 54°
   (b) 65°   (d) 113°

p. 103   Exercise 13

1. 0·6630; 0·9485
2. Each is $\dfrac{\sqrt{3}-1}{2\sqrt{2}}$ {note that $\sin\theta = \cos(90° - \theta)$}
4. 0·8545   6. $2 + \sqrt{3}$
5. 0·8945; − 2   7. 3·0777; 0·5407
9. (1) 0·5592   (2) 0·4848
10. (a) 2·4751   (b) 0·8098

p. 105   Exercise 14

1. 0·96, 0·28, 3·428   6. 0·5
2. 0·4838, 0·8752, 0·5528   8. 0·5; 0·8660
4. 0·9917, − 0·1288   9. 0·6001 approx.
5. (1) 0·9511   (2) 0·3090   12. 0·268 approx.

p. 108   Exercise 15

1. $\frac{1}{2}(\sin 4\theta + \sin 2\theta)$ 9. $2 \sin 3A \cos A$
2. $\frac{1}{2}(\sin 80° − \sin 10°)$ 10. $2 \cos 3A \sin 2A$
3. $\frac{1}{2}(\cos 80° + \cos 20°)$ 11. $2 \sin 3\theta \sin(−\theta)$
4. $\frac{1}{2}(\sin 8\theta − \sin 2\theta)$ 12. $2 \sin 3A \sin 2A$
5. $\frac{1}{2}\{\cos 3(C + D) + \cos(C − D)\}$ 13. $2 \cos 41° \cos 6°$
6. $\frac{1}{2}(1 − \sin 30°) = \frac{1}{4}$ 14. $2 \cos 36° \sin 13°$
7. $\cos 2A − \cos 4A$ 15. $\cot 15°$
8. $\frac{1}{2}(\sin 6C − \sin 10D)$ 16. $\tan \dfrac{\alpha + \beta}{2}$

p. 110   Exercise 16

1. $b = 15·8$; $c = 14·7$ 4. $c = 7·88$; $b = 5·59$
2. $a = 20·3$; $c = 30·4$ 5. $c = 17·3$; $a = 23·1$
3. $a = 7·18$; $c = 6·50$

p. 112   Exercise 17

1. $A = 28° 57'$, $B = 46° 34'$, $C = 104° 29'$
2. $A = 40° 7'$, $B = 57° 54'$, $C = 81° 59'$
3. $A = 62° 11'$, $B = 44° 26'$, $C = 73° 23'$
4. $A = 28° 54'$, $B = 32°$, $C = 119° 6'$
5. $106° 13'$ 6. $43° 51'$

p. 117   Exercise 18

1. $114° 24'$ 2. $29° 52'$ 3. $45° 27'$
4. $A = 22° 18'$, $B = 31° 28'$, $C = 126° 14'$
5. $65°$; $52° 20'$; $62° 40'$ (all approx.)
6. $38° 52'$

p. 120   Exercise 19

1. $A = 25° 30'$; $C = 46° 30'$
2. $A = 64° 19'$; $B = 78° 17'$
3. $B = 99° 46'$; $C = 16° 34'$
4. $83° 25'$; $36° 35'$ 6. $65° 5'$; $42° 41'$
5. $87° 2'$; $63° 44'$

p. 124   Exercise 20

1. $A = 29° 24'$; $B = 41° 44'$; $C = 108° 52'$
2. $A = 51° 19'$; $B = 59° 10'$; $C = 69° 31'$
3. $A = 43° 31'$; $B = 35° 11'$; $C = 100° 18'$
4. $A = 21° 46'$; $B = 45° 27'$; $C = 112° 47'$
5. $A = 35° 23'$; $B = 45° 40'$; $C = 98° 57'$

p. 125   Exercise 21

1. $a = 166·5$; $B = 81° 24'$; $C = 38°$
2. $c = 172$; $A = 32° 42'$; $B = 66° 20'$
3. $b = 65·25$; $A = 33° 26'$; $C = 81° 25'$
4. $c = 286·4$; $A = 65° 18'$; $B = 36° 42'$
5. $b = 136·6$; $A = 58° 38'$; $C = 90° 55'$

p. 127   Exercise 22

1. $b = 145·2$, $c = 60·2$, $B = 81° 28'$
2. $a = 312$, $c = 213$, $C = 42° 41'$
3. $b = 151·4$, $c = 215$, $B = 42° 3'$
4. $a = 152·7$, $b = 83·4$, $A = 97° 41'$
5. $a = 8·27$, $c = 16·59$, $C = 110° 54'$

p. 129   Exercise 23

1. Two solutions : $a = 4·96$ or 58;
                    $A = 126° 4'$ or $3° 56'$
                    $C = 28° 56'$ or $151° 4'$
2. Two solutions : $a = 21·44$ or 109·2
                    $A = 11° 19'$ or $88° 41'$
                    $C = 128° 41'$ or $51° 19'$
3. One solution : $b = 87·08$, $A = 61° 18'$, $B = 52° 42'$
4. Two solutions : $b = 143$ or 15·34.
                    $A = 35°$ or $145°$.
                    $B = 115° 33'$ or $5° 33'$.

p. 131   Exercise 24

1. 19·05 m²   7. 361·3 Mm²
2. 72·36 km²   8. 24·17 m²
3. 39° 25'   9. 0·503 Mg
4. 2537 cm²   10. 239·6 cm²
5. 485 cm²   11. 10 cm
6. 64·8 mm²

p. 132   Exercise 25

1. 5·94 km
2. $A = 88° 4'$, $B = 59° 56'$, $C = 52°$
3. $B = 45° 12'$, $C = 59° 34'$, $a = 726$
4. $C = 56° 6'$
5. 16·35 m, 13·62 m
6. 41°
7. Two triangles: $B = 113° 10'$ or $66° 50'$
                   $C = 16° 50'$ or $63° 10'$
                   $c = 9·45$ or 29·1
8. 267 m approx.   12. 4·5 cm, 6 cm; 11 cm²
9. 6·08 m, 5·71 m   13. 4½ h
10. 3·09 mm   15. 0·3052 m²
11. 7·98 cm, $P = 26° 20'$, 16. 49° 28'; 58° 45'
    $a = 29° 56'$

p. 145   Exercise 26

1. 15·2 m   10. 2·170 km
2. 546 m   11. 500 m approx.
3. 276 m   12. 3·64 km; 45° W. of N.; 5·15 km
4. 193 m approx.   13. 73 m; 51 m
5. 889 m approx.   14. 1246 m approx.
6. 1·26 km   15. 189 m approx.
7. 3700 m   16. 63·7 m approx.
8. 11 990 m   17. 1970 m and 7280 m approx.
9. 2·88 km approx.

p. 152   **Exercise 27**

1. 60°, 15°, 270°, 120°, 135°
2. (a) 0·5878   (c) 0·3090   (e) 0·9659
   (b) 0·9239   (d) 0·3827
3. (a) 4·75   (b) 2·545
4. (a) 13° 24′   (b) 89° 23′
5. (a) $\dfrac{\pi}{12}$ (b) $\dfrac{2\pi}{5}$.   (c) $\dfrac{11\pi}{30}$ (d) $\dfrac{7\pi}{12}$
6. (1) 5·842 cm   (2) 17·5 m
7. $\frac{11}{18}$ radians; 35°
8. 1·57 approx.
9. $\dfrac{\pi}{4}$; $\dfrac{\pi}{3}$; $\dfrac{5\pi}{12}$

p. 165   **Exercise 28**

1. (a) − 0·9744; − 0·2250; 4·3315
   (b) − 0·3619; − 0·9322; 0·3882.
   (c) − 0·7030; 0·7112; − 0·9884
   (d) − 0·2901; 0·9570; − 0·3032
2. (a) − 0·7771   (c) − 0·6691
   (b) 0·7431   (d) − 0·2419
3. (a) − 1·0576   (c) − 1·2349
   (b) 2.   (d) − 1·7434
4. (a) − 0·8387   (c) 1·2799
   (b) 0·7431   (d) 0·5878

p. 173   **Exercise 29**

1. (1) 63°, 117°   (3) 19° 18′, 199° 18′
   (2) 65° 18′, 294° 42′   (4) 65° 6′, 294° 54′
2. (1) 20° 42′, 159° 18′   (2) 18° 26′, 71° 34′
3. (1) 0°, 180°, 80° 32′, 279° 35′
   (2) 43° 52′, 136° 8′
   (3) 45°, 135°, 225°, 315°
   (4) 30°, 150°, 210°, 330°
4. (1) 26° 34′, 45°, 206° 34′, 225°
   (2) 60°, 270°, 300°
   (3) 60°, 300°
   (4) 0°, 120°, 180°, 240°
5. (1) $2n\pi \pm \cos^{-1} 70° 48′$
   (2) $n\pi + (-1)^n \sin^{-1} 19° 42′$
   (3) $n\pi$ or $n\pi + (-1)^n \dfrac{\pi}{6}$
   (4) $n\pi + \dfrac{\pi}{12}$ or $n\pi + \dfrac{5\pi}{12}$
6. (1) 13° 2′   (3) 6° 29′
   (2) 53° 8′   (4) 36° 52′

CALCULUS

P. ABBOTT

This book has been written as a course in calculus both for those who have to study the subject on their own and for use in the classroom.

Although it is assumed that the reader has an understanding of the fundamentals of algebra, trigonometry and geometry, this course has been carefully designed for the beginner, taking him through a carefully graded series of lessons. Progressing from the elementary stages, the student should find that, on working through the course, he will have a sound knowledge of calculus which he can apply to other fields such as engineering.

TEACH YOURSELF BOOKS

THE POCKET CALCULATOR

L. R. CARTER and E. HUZAN

A practical handbook to help you master the techniques of using a pocket calculator.

This book explains calculator applications ranging from the basic arithmetical functions to a wide variety of mathematical, scientific, financial and statistical problems. As such, it is a book both for the beginner and for those wishing to make full use of the facilities available and get the most out of their calculator.

The techniques described are applicable to pocket and programmable calculators alike, and are fully illustrated with examples and exercises (with answers) at varying levels of difficulty.

TEACH YOURSELF BOOKS